W9-BHU-288

SPRINGER
LAB MANUAL

M. Bodanszky A. Bodanszky

The Practice of Peptide Synthesis

Second, Revised Edition

Springer-Verlag
Berlin Heidelberg New York London Paris
Tokyo Hong Kong Barcelona Budapest

Professor Dr. MIKLOS BODANSZKY

One Markham Road
Princeton, NJ 08540, USA

AGNES BODANSZKY (1925–1989)

The First edition was published as Vol. 21 of the series
"Reactivity and Structure Concepts in Organic Chemistry" (3-540-13471-9)

ISBN 3-540-57505-7 Springer-Verlag Berlin Heidelberg New York
ISBN 0-387-57505-7 Springer-Verlag New York Berlin Heidelberg

Library of Congress Cataloging-in-Publication Data. Bodanszky, Miklos. The practice of peptide synthesis / M. Bodanszky, A. Bodanszky.—2nd, rev. ed. p. cm.—(Springer laboratory) Includes bibliographical references and indexes. ISBN 0-387-57505-7 1. Peptides—Synthesis. I. Bodanszky, A. (Agnes) II. Title. III. Series. QD431.B763 1994 547.7′56—dc20 94-890

Printed in Germany

Typesetting: Macmillan India Ltd., Bangalore
51/3130–5 4 3 2 1 0–Printed on acid-free paper

To our daughter Eva

Preface

During the years 1980–81, as guests of the Deutsches Wollforschungsinstitut in Aachen, Germany, we were working on a small book entitled, "Principles of Peptide Synthesis". In the library of the Institute we noted that the volumes of Houben-Weyl's Handbuch der Organischen Chemie dealing with peptide synthesis were so much in use that they were ready to fall apart because the researchers of the Institute consulted them with amazing regularity. They were looking for references, but even more for experimental details which could be adapted to the particular problem they happened to face. In planning a new synthetic endeavor they tried to lean on the experience of others in *analogous* situations. This suggested to us that a smaller and hence more tractable book may be needed, a volume which can be kept on or near the bench to make examples of fundamental methods readily available in the laboratory. Such a collection could save numerous short trips to the library, a point particularly important where a library well equipped with the sources of the literature of peptide synthesis is not near at hand. Also, we thought that the envisaged book may be welcome by those who are more versed in English than in German. To the best of our knowledge no similar publication is available.

In our attempt to provide the peptide chemist with a collection of well established procedures we resisted the temptation to include novel, but untested methods, although some of these are quite original and hence intellectually attractive. It seems, that in the practical execution of peptide synthesis most investigators stick to classical approaches. It might indeed be somewhat imprudent to add to the risks inherent in the synthesis of a long chain potentially hidden unknown factors of yet untried methods. In certain respects the synthesis of peptides is different from the synthesis of other organic compounds. The construction of a longer peptide chain involves the simultaneous handling of numerous functional groups and this difficulty is often compounded by the high molecular weight of the target compounds. The ensuing technical problems, such as the lack of solubility in the commonly used solvents, the need to carry out bimolecular reactions at low concentrations of the reactants and, last but not least, limitations in the analytical information available during synthesis, all warn the investigator to be cautious. Thus, most peptide chemists are conservative in their choice of approach and probably rightly so.

Several examples selected for demonstration of well established methods result in compounds which are, by now, commercially available. Yet, the work of the peptide chemist may require materials which are not available from research supply houses, for instance derivatives of unusual amino acids. In such a case, the synthesis can follow the pattern used in the preparation of a trivial intermediate. Also, the methods described for small peptides are usually applicable, *mutatis mutandis*, in operations involving more complex materials as well. The examples assembled in this book should serve in preparative studies which are based on *analogies*. To render these examples more practical we followed the literature faithfully but not exactly. Amounts of starting materials were recalculated and expressed in molar ratios. The quantities of liquids are given both in weight and in volume. More importantly, points which require comment are followed by footnotes and in these an attempt was made to put the extensive experience of the authors to good use. Hazards were pointed out to protect the uninitiated and pertinent references were added to assist those who wish to look deeper into the matter. We hope that our colleagues active in the field of peptide synthesis will consider "The practice of Peptide Synthesis" a worthwhile addition to the rich literature of peptide chemistry.

Cleveland, Ohio
February, 1984

MIKLOS BODANSZKY
AGNES BODANSZKY

Preface to the Second Edition

Continued interest in The Practice of Peptide Synthesis prompted the publication of this Second Edition. Revision of the First Edition was not limited to correction of errors. The book now contains several procedures, which have gained significance during recent years; for instance, additional blocking groups of the thiol function and novel coupling reagents. The author hopes that these changes enhance the usefulness of the volume for investigators active in the field of peptide synthesis.

Princeton, New Jersey
May, 1994

M.B.

Contents

I Introduction

In the selection of examples of well established procedures a series of decisions had to be made in order to keep the material to be presented within practical limits. These decisions resulted in a seemingly distorted picture of peptide synthesis. The number of procedures for the introduction of protecting groups appears to be excessive or at least not commensurate with the number of methods applicable for their removal. Also, relatively few methods of coupling were rendered and still less coupling reagents. These choices, however, were made not without good reasons. There is indeed a definite need for a plethora of blocking groups. The simultaneous handling of numerous side chain functions in combination with various methods of coupling and with potential side reactions related to a particular sequence of amino acids is possible only if a whole gamut of masking groups is available. For the removal of protecting groups many proposals can be found in the literature, but a closer scrutiny reveals that most of these are variations on a few themes: reduction, e.g. hydrogenolysis, acidolysis and displacement by nucleophiles. A similar repetitiveness can be discerned among coupling methods. Only exceptionally can one find methods for the formation of the peptide bond, which are not based on symmetrical or mixed anhydrides or on the aminolysis of reactive esters. Therefore, we confined ourselves mainly to the presentation of the few principal procedures. Coupling reagents deserve a special mention at this point. There seems to exist a certain fascination in connection with coupling reagents. The idea of adding some magic compound to the mixture of a carboxylic acid and an amine and thereby accomplishing the formation of an amide bond attracted many investigators. Yet, most coupling reagents simply cause the activation of a carboxyl group by converting it into an anhydride or an active ester. It is not obvious, however, why the carboxyl should be activated in the presence rather than in the absence of the amino component. Furthermore, a true coupling reagent should be completely inert toward amines. Unfortunately very few of the reactive materials proposed for this purpose have this necessary attribute. Carbodiimides stand out in this respect: the rate of their reaction with amines (to form guanidine derivatives) is usually negligible. This accounts for their undiminished popularity although more than a quarter of a century has passed since their introduction [1] in peptide synthesis. The use of carbodiimides, particularly in the presence of auxiliary nucleophiles [2] is, at this time, one of the most important approaches to peptide bond formation and had to be demonstrated in more than one example, but for the reasons just outlined, very few other coupling reagents are discussed.

So far no well-established process is available for the activation of the amino component and peptide bond formation is generally accomplished through reactive derivatives of carboxylic acids. This requires activation of the carboxyl group with more or less powerful reagents, such as alkyl chlorocarbonates, carbodiimides, etc. In more recent years the long standing desire [3] to avoid such aggressive chemicals in the preparation of peptides and to follow Nature in the use of enzyme-catalyzed reversible reactions led to the practical application of proteolytic enzymes for the formation of the peptide bond. Major progress has been made in this direction by the selection of suitable enzymes and pH ranges which are more favorable for the synthesis than for the hydrolysis of the amide bond. A most important contribution in this area is the addition of organic solvents [4] to the reaction mixture. To indicate the growing significance of this approach we include a few examples of enzyme-catalyzed syntheses and also some enzymatic methods of removal of protecting groups. Similarly, the semisynthesis of peptides and proteins, the construction of large molecules from fragments of proteins, will be treated only briefly; a more detailed presentation would transcend the limits of this volume.

We had to give special consideration to techniques of facilitation such as the "handle" method [5], to synthesis of peptides attached to soluble polymers (or "liquid phase" peptide synthesis [6]), to the "in situ" technique [7] and particularly to the extremely popular method of solid phase peptide synthesis [8]. It seemed impractical to add examples of these promising and already very significant approaches in a number commensurate with the available literature. Solid phase peptide synthesis itself requires a separate volume for proper presentation. The execution of solid phase synthesis has been rendered by Stewart and Young and an updated version of their book became available [9]. The extensive literature was assembled in review articles and books [10, 11] and also in a "user's guide" for the preparation of synthetic peptides by the solid-phase method [12]. We still deemed it necessary to include a few examples of syntheses in which techniques of facilitation were applied in the hope that the examples are sufficiently representative and will be found useful by the readers.

The word "technique" in connection with peptide synthesis calls to mind also some simple technical aspects of preparative work in organic synthesis. We mention here a seemingly trivial example, the separation of a solid intermediate from the solvent and from the by-products in solution. Only thoughtfully designed and carefully executed techniques of filtration provide an intermediate which can be used without purification in the next step of the chain building procedure. If the filter cake is not properly packed down, e.g. with the help of a sturdy glass rod with a flattened head, then it is not likely that the impurities are completely removed by displacement with a limited volume of the solvent (judiciously) selected for washing. The authors apologize for this somewhat lengthy discussion of such minor points as a glass rod with a flattened head or a sinter-glass funnel. We are convinced, however, that the difference between success and failure in synthesis can hinge on such imponderabilia. This is particularly true when in stepwise chain building poor solubility of an intermediate in the commonly used organic solvents (including dimethylformamide) prevents extensive purification by chromatography, electrophoresis or countercurrent distribution. Also, even when these efficient methods of purification are applicable, they are time consuming. If they are avoidable simply by washing of the intermediates with appropriate solvents, the tedium of synthesis is greatly reduced. For such reasons similar technical aspects are the subjects of comments throughout the book.

Other comments refer to health hazards. Operations with noxious materials such as liquid ammonia, liquid HF or HBr in acetic acid obviously demand a well ventilated hood. Enthusiasm about or dedication to our objectives can make us forgetful of the dangers surrounding our work. Therefore it may not be superfluous to add reminders which call the attention to some potential harm. This does not, however, absolve the researcher from the usual caution which belongs to the practice of organic chemistry, like the protection of the eyes, hands or protection of our colleagues in the laboratory.

In the ideal synthesis a single peptide is produced which requires no purification. Even if this objective is usually not achieved, every attempt has to be made to avoid the other extreme in which the target compound must be "fished out" from a mixture of numerous closely related peptides. Situations can (and not seldom did) arise where the peptidic material prepared by synthesis is an intractable mixture and the endeavor already progressed to a late stage has to be abandoned. The best way to prevent this is to select only methods which, at least in principle, give rise to a single product and to execute all operations in an unequivocal manner. Thus, the investigator must know the possible alternative pathways the reactants might follow. He must be well informed about possible side reactions. Reviews [13] on side reactions in peptide synthesis tried to provide some help in this respect, but the peptide chemist has to remember Murphy's law: whatever can go wrong, will. Such dangers, however, should not be considered deterrents: many already experienced side reactions notwithstanding, fairly complex peptides have been secured in high yield and in good quality by competent peptide chemists.

In addition to chemical factors which need to be considered in peptide synthesis, some of the physical properties of synthetic peptides can also cause difficulties. Peptides are often polyelectrolytes and if they have several cationic centers they can be adsorbed on glass, a polyanion. Losses of peptides were noted, particularly at high dilution, for instance in solutions used in pharmacological assays. The hydrophobic regions in peptide chains can similarly lend themselves to adsorption, namely on polyethylene. A related and sometimes major problem is created by the self-association (aggregation) of peptides, the formation of insoluble particles and the consequent loss of valuable material. No simple remedies can be offered for these potential losses, but awareness should lead to solution of such problems; for instance, absorption on glass can be prevented by presaturation of the surface with polycations (e.g., serum albumin).

A few closing sentences must be dedicated to the analytical control of peptide synthesis. Because of the high molecular weight and complex structure of most biologically active peptides, their analysis is more problematic than the analysis of many other products of organic synthesis. Thus, the execution of elemental analysis is hampered by the tendency of peptides to tenaciously retain water or solvents such as acetic acid. The ionic character of peptides causes further complications in this respect: weak cationic centers, such as the nitroguanidino group or the imidazole nucleus, can remain associated with acids, e.g. trifluoroacetic acid used in preceding steps. In spite of such complications the practitioners of peptide synthesis should not abandon elemental analysis. Prolonged drying in good vacuum at elevated temperature might be necessary and the dried sample may require special handling, as if it were hygroscopic. The carefully prepared material, however, should give satisfactory values for the elements which constitute the peptide. Such a satisfactory analysis should be considered necessary but not sufficient evidence for the homogeneity of the synthetic product. Amino acid analysis of a carefully hydrolyzed sample will provide additional valuable information which can be further supplemented by u.v. spectra, if the peptide contains residues with chromophores in the side chain (tyrosine, tryptophan or nitroarginine). Infrared spectra are usually less informative, but in special cases, such as in peptides in which the tyrosine residue is esterified with sulfuric acid, the evidence provided by i.r. spectra is quite important. Infrared spectroscopy is more useful in the examination of starting materials, for instance active esters, anhydrides, insoluble polymeric supports. The value of nmr spectra cannot be overestimated, but again, they are less informative when used for large molecules. An important but not always available tool of analysis is the sequencing of intermediates or, even more importantly, of the final product. The synthetic intermediates can and, whenever possible, should be scrutinized on thin layer chromatograms and by high pressure liquid chromatography. The complexity of the products of peptide synthesis demands that not one analytical method should be used for their examination but as many as possible.

The authors attempted here to point out some of the more important factors which influence the outcome of an endeavor toward a synthetic peptide. Yet, like probably all human undertakings, peptide synthesis also requires some good luck, and this is what we wish our colleagues embarking on demanding ventures.

1. Sheehan JC, Hess GP (1955) J Am Chem Soc 77: 1067
2. König W, Geiger R (1970) Chem Ber 103: 1067
3. Waldschmidt-Leitz E, Kühn K (1957) Chem Ber 84: 381
4. Homandberg M, Mattis JA, Laskowski M, Jr. (1978) Biochemistry 17: 5220
5. Camble R, Garner R, Young GT (1968) Nature 217: 247
6. Mutter M, Bayer E (1972) Nature 237: 512
7. Bodanszky M, Funk KW, Fink ML (1973) J Org Chem 38: 3565; Bodanszky M, Kondo M, Yang-Lin C, Siegler GF (1974) ibid 39: 444
8. Merrifield RB (1963) J Amer Chem Soc 85: 2149
9. Stewart JM, Young JD (1984) Solid Phase Peptide Synthesis, Second Edition, Freeman, San Francisco, CA
10. Birr C (1978) Aspects of the Merrifield Peptide Synthesis, Springer Verlag, Berlin, Heidelberg New York
11. Atherton E, Sheppard RC (1989) Solid Phase Peptide Synthesis, A Practical Approach, IRL Press at Oxford University Press, Oxford
12. Grant GA (1992) Synthetic Peptides. A User's Guide, Freeman, New York
13. (Former Ref 12) Bodanszky M, Martinez J (1983) In: Gross E, Meienhofer J (eds) The Peptides, Academic Press, New York, Vol 5, p 111

II Protecting Groups

1 Introduction of Amine Protecting Groups

1.1 The *p*-Toluenesulfonyl Group [1]

p-Toluene-sulfonyl-L-isoleucine [2]

$$CH_3-C_6H_4-SO_2Cl + H_2N-CH(CH(CH_3)CH_2CH_3)-COONa \xrightarrow[2.\ HCl]{1.\ NaOH} CH_3-C_6H_4-SO_2-NH-CH(CH(CH_3)CH_2CH_3)-COOH$$

$C_{13}H_{19}NO_4S$ (285.4)

L-Isoleucine (13.1 g, 100 mmol) is suspended (and partially dissolved) in distilled water (50 ml) and completely dissolved by the addition of N NaOH (100 ml). *p*-Toluenesulfonyl chloride (26.7 g, 140 mmol) is added to the vigorously stirred solution followed by N NaOH in small portions to maintain the alkalinity of the mixture at about pH 9. Some heat is evolved during the reaction and external cooling with cold water or ice-water is necessary to keep the temperature of the reaction mixture at about 20 °C. After no more alkali is used up [3] stirring is continued at room temperature one hour longer. Any unreacted acid chloride is removed by filtration and the solution acidified with 5 N HCl (about 20 ml) to Congo. The mixture is stored in the cold overnight; the crystals are collected on a filter, washed with water, dried in air and finally in vacuo over P_2O_5. The product, 25 g (88%) melts at 135–136 °C; $[\alpha]_D^{21}$ $-12°$ (*c* 2, 0.5 N $KHCO_3$) [4, 5].

1. Fischer E, Livschitz W (1915) Ber dtsch Chem Ges 48: 360.
2. Katsoyannis, PG, du Vigneaud V (1954) J Amer Chem Soc 76: 3113

3. In addition to the 100 ml of N NaOH used for the formation of the sodium salt of isoleucine about 140 ml more N NaOH is needed. It is added over a period of 2 hours.
4. For the m.p. of several other tosylamino acids cf. McChesney EW, Swan WK, Jr. (1937) J Amer Chem Soc 59: 1116
5. In the preparation of the *p*-toluenesulfonyl derivatives of phenylalanine and tyrosine the poorly soluble sodium salt of the tosyl-derivative separates during the reaction. The suspension is acidified and the product extracted into ether from which it separates in crystalline form.

1.2 Phthalyl(Phthaloyl, Pht) Amino Acids

Phthalyl-L-leucine [1, 2]

$C_{14}H_{15}NO_4$ (261.3)

A vigorously stirred solution of L-leucine (13.2 g, 100 mmol) and sodium carbonate decahydrate (29 g, 101 mmol) in distilled water (150 ml) is treated with finely powdered *N*-ethyloxycarbonylphthalimide [3] (23 g, 105 mmol). In about 15 minutes almost all the reagent dissolves. The solution is filtered from a small amount of unreacted material and acidified to Congo [4] with 6 N HCl (about 34 ml). The precipitated phthalyl-L-leucine is collected on a filter, washed with water and dried in air. It can be recrystallized from toluene-hexane. The product, 24.3 g (93%) melts at 100 °C; $[\alpha]_D^{25}$ −25° (*c* 2, 95% ethanol).

1. Nefkens HGL, Tesser GI, Nivard RJF (1960) Recueil 79: 688. The preparation of several additional phthalylamino acids is described. Phthalyl-L-tryptophan could not be obtained by this method.
2. Phthalyl amino acids obtained by the fusion of phthalic anhydride with amino acids (Billman JH, Harting WF (1948) J Amer Chem Soc 70: 1473) are not suitable for peptide synthesis because their chiral purity is lost in the process. Several alternative methods have been proposed for phthalylation of amino acids but none of these can compete in simplicity with the procedure of Nefkens et al. (ref. 1).
3. The reagent is commercially available. If necessary, it can be prepared from phthalimide (147 g, 1 mol) which is dissolved in dimethylformamide (0.5 liter) by the addition of triethylamine (101 g = 140 ml, 1 mol). The solution is cooled in an ice-water bath while slowly ethyl chlorocarbonate (113.5 g = 100 ml, 1.05 mol) is added with stirring. About an hour is required for the addition of the chlorocarbonate. Stirring is continued for about another hour while the mixture is allowed to warm up to room temperature. The solution is poured into water (3 liters) with vigorous stirring. The product is collected on a filter, thoroughly washed with water, dried in air and finally in vacuo over phosphorus pentoxide. The crude reagent (190 g, 86%) is purified by recrystallization from hot ethanol to reach the m.p. 80 °C. Two recrystallizations might be necessary (In the literature a m.p. of 87–89 °C is also reported: Heller G, Jacobsohn P (1921) Ber dtsch Chem Ges 54: 1107)
4. The pH of the suspension should be between 2 and 3.

1.3 The Benzyloxycarbonyl Group [1, 2]

Benzyloxycarbonyl-L-proline [3]

$$C_6H_5-CH_2O-CO-Cl + HN-CH-COONa \text{ (pyrrolidine ring)} + NaOH \longrightarrow$$

$$C_6H_5-CH_2O-CO-N-CH-COONa \xrightarrow{HCl} C_6H_5-CH_2O-CO-N-CH-COOH + NaCl$$

$C_{13}H_{15}NO_4$ (249.3)

A solution of L-proline (115.1 g, 1.0 mol) in 2 N NaOH (500 ml) is cooled in an ice-water bath and stirred with a powerful magnetic stirrer. Benzyl chlorocarbonate [4] (187 g = 158 ml, 1.1 mol) and 2 N NaOH (550 ml) are added in about ten portions, alternatingly. The reaction of the mixture should remain distinctly alkaline: if necessary more 2 N NaOH is added. The temperature of the reaction mixture is kept between +5 and +10 °C by the rate of addition of the reactants. About one to one and a half hour is needed. The ice-water bath is replaced then by water of room temperature and vigorous stirring continued for an additional 30 min. The alkaline solution is extracted four times with ether (0.5 liter each); the ether extracts are discarded [5]. The ether dissolved in the aqueous layer is removed by bubbling a steam of nitrogen through the solution [6] which is then acidified to Congo blue by the addition of 5 N HCl (about 220 ml). An oil separates and slowly solidifies [7] to a crystalline mass. This is disintegrated to a powder, filtered, washed with distilled water (0.5–1.0 liter) and dried in air. The product, 230 g (92%) melts at 75–77 °C, $[\alpha]_D^{20}$ −61° (*c* 2 to 5, AcOH). This material is sufficiently pure for most practical purposes. It can be recrystallized from carbon tetrachloride or from ethyl acetate-hexane.

Benzyloxycarbonyl-L-leucine

$$C_6H_5-CH_2O-CO-Cl + H_2N-CH(CH_2-CH(CH_3)_2)-COONa \xrightarrow{NaOH} C_6H_5-CH_2O-CO-NH-CH(CH_2-CH(CH_3)_2)-COONa$$

$$\xrightarrow{HCl} C_6H_5-CH_2O-CO-NH-CH(CH_2-CH(CH_3)_2)-COOH + NaCl$$

$C_{14}H_{19}NO_4$ (265.3)

To a solution of L-leucine (1.0 mole, 131.1 g) in water (300 ml) 5 N NaOH (200 ml) is added and the stirred solution is cooled in an ice-water bath. Benzyl chlorocarbonate [4] (187 g = 158 ml, 1.1 mol) and 2 N NaOH (550 ml) are added in ten portions, alternatingly, while the mixture is vigorously stirred and its temperature maintained at about 10 °C. The additions are completed in about one and a half hours. After continued stirring at room temperature for 30 min the alkalinity of the mixture is adjusted to about pH 10 and the solution is extracted with ether four times (0.5 liter each) [5]. The aqueous layer is acidified to Congo blue with 5 N HCl (about 200 ml) and the oil which separates extracted into ether (about 1.2 liter, in 3 portions). The ether solution is dried over $MgSO_4$ and the solvent removed in vacuo. The residue, a syrup, is taken up in toluene (1 liter) and the solution is concentrated, in vacuo, to about 1 liter. This stock-solution can be stored and aliquots used as needed [8]. The benzyloxycarbonyl-L-leucine content of the solution is determined by titrating a 1.0 ml aliquot, after dilution with a few ml of 95% ethanol, with 0.1 N NaOH in the presence of phenolphthaleine. The yield is about 240 g (90%).

N^{α}-Benzyloxycarbonyl-L-arginine [9]

C_6H_5–CH_2O–CO–Cl + H_2N–CH(CH_2–CH_2–CH_2–NH–C(=NH)–NH_2)–COONa ⟶ C_6H_5–CH_2O–CO–NH–CH(CH_2–CH_2–CH_2–NH–C(=NH_2^+)–NH_2)–COO^-

$C_{14}H_{20}N_4O_4$ (308.3)

Arginine monohydrochloride (21.3 g, 100 mmol) is dissolved in ice-cold N NaOH (100 ml) with stirring. Both stirring and cooling to about 0 °C are continued while benzyl chlorocarbonate (22.1 g = 18.5 ml, 130 mmol) and 2 N NaOH (55 ml) are added, in a few portions, alternatingly. The pH of the mixture is kept between 9 and 10. After the addition of the reactants is completed stirring of the suspension is continued for two more hours. The pH drops during this time to 7–7.5. The precipitate is collected on a filter, washed with cold (ca. 10 °C) water (50 ml) and recrystallized from boiling water (about 130 ml). Crystallization is completed in the cold (ice-water bath). The product is collected, dried in air and then suspended (in finely powdered form) in acetone (50 ml) filtered, washed with acetone (20 ml) and with ether (50 ml). The purified material is dried in vacuo at 50 °C. It weighs 27.6 g (89.5%), melts at 184 °C dec. [10]; $[\alpha]_D^{23}$ −9.3° (*c* 2, N HCl). Elemental analysis gives correct values for C, H and N.

Benzyloxycarbonyl-L-aspartic Acid β-Benzyl Ester [11] (β-Benzyl Benzyloxycarbonyl-L-aspartate)

C_6H_5–CH_2O–COCl + H_2N–CH(CH_2–CO–OCH_2–C_6H_5)–COONa —(1. $NaHCO_3$; 2. HCl)⟶ C_6H_5–CH_2O–CO–NH–CH(CH_2–CO–OCH_2–C_6H_5)–COOH

$C_{19}H_{19}NO_6$ (357.4)

L-Aspartic acid β-benzyl ester [11] (2.23 g, 10 mmol) is dissolved in hot distilled water (150 ml). The solution is allowed to cool to 60 °C [12], treated with $NaHCO_3$ (2.1 g, 25 mmol) and benzyl chlorocarbonate (2.1 g = 1.8 ml, ca 12 mmol) and allowed to cool slowly to room temperature under vigorous stirring. Stirring is continued for about 3 hours. The solution is extracted with ether (twice, 75 ml each time) and acidified to Congo with concentrated hydrochloric acid. The product separates either as a solid or as an oil, which soon solidifies. After several hours in the refrigerator the disintegrated material is collected on a filter, washed with water and dried at 50 °C in vacuo. After recrystallization from benzene [13] benzyloxycarbonyl-β-benzyl-L-aspartic acid (2.65 g, 75%) melts at 107–108 °C, $[\alpha]_D^{25}$ +11.9° (*c* 10, AcOH).

1. Bergmann M, Zervas L (1932) Ber dtsch Chem Ges 65: 1192
2. The earlier used nomenclature, carbobenzoxy or carbobenzyloxy, has been replaced by benzyloxycarbonyl and the abbreviation "cbz" by the capital letter "Z".
3. The preparation of Z-Pro has been described by several authors, cf. e.g. Berger A, Kurtz J, Katchalski E (1954) J Am Chem Soc 76: 5552
4. Benzyl chlorocarbonate is commercially available, often under the name benzyl chloroformate or carbobenzoxy chloride. It can deteriorate on storage, probably due to disproportionation to dibenzyl carbonate and phosgene. It is advisable, therefore, to bubble nitrogen through the sample to be used, in a hood, for several hours. The benzyl chlorocarbonate content of the material can be determined in a small scale experiment, e.g. by the acylation of excess glycine. For the introduction of the benzyloxycarbonyl group into unusual amino acids or into valuable peptides, freshly prepared benzyl chlorocarbonate may be preferable to commercial materials. For its preparation cf. Farthing AC, J Chem Soc 1950: 3213 or Organic Syntheses Vol. 23, p. 13. Benzyl chlorocarbonate can be distilled but only at moderate temperature in high vacuum.
5. The ether extracts should not be poured into a sink or down the drain. They can be treated with a small volume of ammonium hydroxide solution and allowed to evaporate in a hood, far from any possible source of fire.
6. The ether vapors should be removed in a hood.
7. Crystallization can be very slow. In fact, benzyloxycarbonyl-L-proline has been known for many years as an oil. Seeding greatly accelerates the process. Seed crystals can be obtained by crystallization of a small sample from carbon tetrachloride, from ether-hexane or from ethylacetate-hexane.
8. In the preparation of benzyloxycarbonyl-L-phenylalanine according to the procedure described here for the preparation of benzyloxycarbonyl-L-leucine, acidification of the alkaline reaction mixture generally yields a solid, which is the mixture of the desired product and of an adduct between benzyloxycarbonyl-L-phenylalanine and its sodium salt. Therefore, the process of acidification should be reversed by pouring the alkaline solution in a thin stream into a small excess of N HCl, stirred and cooled with small pieces of ice.
   As far as we know, benzyloxycarbonyl-L-leucine has not been obtained so far in crystalline form. Since the syrupy material tenaciously retains solvents, including ether, the exact determination of the amount used for a particular synthesis is not easy. It seems to be practical to use a stock solution as described above. Such solutions can be stored for prolonged periods, even for years.
9. Boissonnas RA, Guttmann S, Huguenin RL, Jaquenoud PA, Sandrin E (1958) Helv Chim Acta 41: 1867
10. Originally (Bergmann M, Zervas L (1932) Ber dtsch Chem Ges 65: 1192) a m.p. of 175 °C was reported.

11. Benoiton L (1962) Can J Chem 40: 570; cf. also Berger A, Katchalski E (1951) J Amer Chem Soc 73: 4084
12. The starting material starts to separate at 55 °C.
13. Benzene is quite toxic and probably an alternative solvent should be sought, but purification of the product is indeed advisable. If strong alkali, commonly applied in the acylation of amino acids, has to be replaced by bicarbonate (as in the case of amino acid esters or amides or in the introduction of base-labile blocking groups), the product is usually contaminated by di- and tripeptide derivatives, generated via mixed anhydrides.

1.4 4-Methoxybenzyloxycarbonylamino Acids [1, 2]

4-Methoxybenzyl Chlorocarbonate

$$CH_3O-C_6H_4-CH_2OH + COCl_2 \longrightarrow CH_3O-C_6H_4-CH_2O-CO-Cl + HCl$$

A solution of phosgene (100 g, one mol) in dry ether (500 ml) is cooled in an ice-water bath while 4-methoxybenzyl alcohol (69 g, 0.5 mol) is added, dropwise, with stirring. About 30 min is required for the addition of the alcohol. The ether is removed in vacuo at a bath temperature of 0 °C or below [3] until the volume of the solution is reduced to about 200 ml. This solution of 4-methoxybenzyl chlorocarbonate is used for the preparation of 4-methoxybenzyloxycarbonylamino acids [4].

Introduction of the Protecting Group

$$CH_3O-C_6H_4-CH_2O-CO-Cl + H_2N-CHR-COONa \xrightarrow[2.\ HCl]{1.\ NaOH} CH_3O-C_6H_4-CH_2O-CO-NH-CHR-COOH$$

A vigorously stirred solution of the amino acid (100 mmol) in 2 N NaOH (50 ml) and tetrahydrofuran (50 ml) is cooled in an ice-water bath and treated with an aliquot (corresponding to 125 mmol 4-methoxybenzyl alcohol) of the ethereal solution of 4-methoxybenzyl chlorocarbonate described above and 2 N NaOH (ca 62 ml). The two solutions, that of the chlorocarbonate and the 2 N NaOH, are added alternately, each in about 5 portions. About 30 min is needed for the addition of the reagents. The pH of the reaction mixture is maintained at 9 to 10, if necessary with more 2 N NaOH. The ice-water in the bath is replaced by water of room temperature and stirring is continued for one hour. The pH is adjusted to 7 with NaOH or with acetic acid and the tetrahydrofuran is removed in vacuo. The remaining aqueous solution is extracted with ether (100 ml) and acidified to pH 2 with 2 N HCl. The protected amino acid, usually an oil [5] is extracted into ethyl acetate (200 ml), the extract washed with water and dried over $MgSO_4$. The solvent is removed in vacuo and the residue purified by recrystallization [6].

1. Sakakibara S, Honda I, Naruse M, Kanaoka M (1969) Experientia 25: 576
2. Several alternative methods have been proposed for the preparation of 4-methoxybenzyloxycarbonylamino acids, for instance: a) Weygand F, Hunger K (1962) Chem Ber 95: 1; b)

Jones JH, Young GT: Chem Ind 1966: 1722; c) Klieger E (1969) Justus Liebigs Ann Chem 724: 204; d) Sofuku S, Mizumura M, Hagitani A (1970) Bull Chem Soc Japan 43: 177; e) Yajima H, Tamura F, Kiso Y, Kurobe M (1973) Chem Pharm Bull 21: 1380. The procedure of Sakakibara et al. (ref. 1) was selected for this volume because of its simplicity.

3. An ice-salt bath was recommended by Sakakibara et al. (ref. 1); Jones and Young (ref. 2c) applied an ice-water bath during evaporation.
4. The chlorocarbonate is not sufficiently stable for prolonged storage. In ethereal solution it can be stored in a deep freezer for a few days, but decomposition takes place in solutions more concentrated than 2.5 molar or if chloroform is added to the solution. Therefore, for the preparation of 4-methoxybenzyloxycarbonylamino acids on a smaller scale it might be preferable to apply the azide procedure (ref. 2a) or to use a mixed carbonate (refs. 2a, 2b, 2c, 2d).
5. The derivatives of glutamine and of asparagine separate as solids at this point. These are collected on a filter, washed with water and dried.
6. In most cases crystals were obtained from ethylacetate-hexane or from abs. ethanol-hexane. Physical properties of a series of 4-methoxybenzyloxycarbonylamino acids are described in ref. 1.

1.5 The *tert*-Butyloxycarbonyl (Boc) Group

1.5.1 Introduction of the *tert*-Butyloxycarbonyl Group with 2-*tert*-Butyloxycarbonyloximino-2-phenylacetonitrile [1, 2]

2. *tert*-Butyloxycarbonyloximino-2-phenylacetonitrile [2, 3]

$C_6H_5-CH_2-CN + CH_3-O-NO \longrightarrow C_6H_5-C(=NOH)-CN \xrightarrow[CCl_3O-CO-Cl]{COCl_2 \text{ or}}$

$C_6H_5-C(=N-O-CO-Cl)-CN \xrightarrow[\text{pyridine}]{(CH_3)_3COH} C_6H_5-C(=N-O-CO-O-C(CH_3)_3)-CN$

$C_{13}H_{14}N_2O_3$ (246.3)

A solution of sodium hydroxide (40 g, 1 mol) in methanol (300 ml) is cooled in an ice-water bath and benzyl cyanide (117 g = 116 ml, 1 mol) is added with stirring. Stirring and cooling are continued while gaseous methyl nitrite is introduced. The latter is generated from a suspension of $NaNO_2$ (83 g, 1.2 mol) in a mixture of methanol (53 ml) and water (50 ml) by the dropwise addition of a mixture of concentrated sulfuric acid (32 ml) and water (65 ml) [4]. The reaction mixture is stirred at room temperature for two hours and evaporated to dryness in vacuo. The residue is dissolved in water, extracted twice with toluene and the aqueous layer acidified (to Congo) with concentrated hydrochloric acid. The precipitated oxime is collected by filtration, washed with water and dried in air: 120 g (82%), m.p. 119–124 °C, R_f 0.50 (in chloroform-methanol 9:1). The crude material (14.6 g, 100 mmol) and *N*,*N*-dimethylaniline (12.1 g = 13 ml, 100 mmol) are dissolved in a mixture of dioxane (5 ml) and benzene (100 ml) and are added dropwise to a stirred and cooled solution of trichloromethyl chloroformate (10.9 g = 6.7 ml, 55 mmol) or 10.9 g (110 mmol) phosgene [5] in benzene (30 ml) at 3–5 °C. The mixture is stirred at room temperature for 6 hours and stored overnight. Next day a mixture of *tert*-butanol (11.1 g = 14.1 ml, 150 mmol) and pyridine (15.6 g = 16 ml, 197 mmol) in benzene (20 ml) is added at 5–10 °C. After stirring for 3 hours at this temperature and 4 hours at room temperature the mixture is allowed to stand at room temperature overnight. The solution is then extracted with water, N HCl, water, 5% $NaHCO_3$ solution and again with water. It is dried over $MgSO_4$ and evaporated in vacuo. The residue is triturated with aqueous (90%) methanol (20 ml), filtered, washed with the same solvent mixture (30 ml) and dried: 17.0 g (69%), m.p. 84–86 °C; R_f 0.74 (in chloroform-methanol 9:1). The i.r. spectrum shows a carbonyl band at 1785 cm^{-1}. In the nmr spectrum ($CDCl_3$) the 9 methyl protons appear as a singlet at 1.62 ppm, the aromatic protons as a multiplet between 7.8 and 8.2 ppm. Correct analytical values are obtained for C, H and N.

tert-Butyloxycarbonylglycine [2]

$$C_6H_5-C(CN)=N-O-CO-O-C(CH_3)_3 + H_2N-CH_2-COOH \xrightarrow[2.\ H^+]{1.\ N(C_2H_5)_3} (CH_3)_3C-O-CO-NH-CH_2-COOH + C_6H_5-C(CN)=NOH$$

$C_7H_{13}NO_4$ (175.2)

The reagent (2-*tert*-butyloxycarbonyloximino-2-phenylacetonitrile, 27.1 g, 110 mmol) is added as a finely powdered solid to a stirred solution of glycine (7.5 g, 100 mmol) and triethylamine (15.1 g = 21 ml, 150 mmol) in a mixture of dioxane (60 ml) and water (60 ml). Stirring is continued at room temperature for 2 hours. Water (150 ml) and ethyl acetate (200 ml) are added, the aqueous phase reextracted with ethyl acetate (200 ml) and acidified to Congo with a 5% solution of citric acid in water. The precipitate is collected, washed with water and dried in air: 15.2 g, (87%), m.p. 86.5–87.5 °C.

tert-Butyloxycarbonyl-L-tryptophan [2]

Finely powdered 2-*tert*-butyloxacarbonyloximino-2-phenylacetonitrile (27.1 g, 110 mmol) is added to stirred solution of L-tryptophan (20.4 g, 100 mmol) and triethylamine (15.1 g = 21 ml, 150 mmol) in water (60 ml) and dioxane (60 ml). After about one hour the mixture becomes homogeneous; stirring is continued for two more hours. Water (150 ml) and ethyl acetate (200 ml) are added, the aqueous layer separated, reextracted with ethyl acetate (200 ml) and then acidified to Congo with a 5% solution of citric acid in water. The product is extracted with ethyl acetate (twice, 200 ml each time), the ethyl acetate extracts pooled, washed with water (twice, 100 ml each time), dried over $MgSO_4$ and evaporated in vacuo to dryness. The crystalline material [6] (30 g, 98.6%) melts at 137–138 °C with decomposition.

1. Itoh M, Hagiwara D, Kamiya T, Tetrahedron Letters 1975: 4393
2. Itoh M, Hagiwara D, Kamiya T (1977) Bull. Chem. Soc. Japan 50: 718
3. This reagent is commercially available.
4. Methyl nitrite is highly toxic. The operations must be carried out under a well ventilated hood.
5. Both phosgene and its dimer are extremely toxic. A good hood is required.
6. Boc-amino acids, which are difficult to obtain in crystalline form, can be converted to the stable and generally crystalline dicyclohexylammonium salts by adding dicyclohexylamine (1.81 g = 2.0 ml, 10 mmol) to a solution of the Boc-amino acid (10 mmol) in ether.

1.5.2 Introduction of the *tert*-Butyloxycarbonyl Group with *tert*-Butyl Pyrocarbonate [1–3]

Boc-Amino Acids

$$CH_3-C(CH_3)_2-O-CO-O-CO-O-C(CH_3)_2-CH_3 + H_2N-CHR-COOH \longrightarrow CH_3-C(CH_3)_2-O-CO-NH-CHR-COOH + CO_2 + HO-C(CH_3)_2-CH_3$$

A solution of the amino acid (10 mmol) in a mixture of dioxane (20 ml), water (10 ml) and 1 N NaOH [4] (10 ml) is stirred and cooled in an ice-water bath. Di-tert-butyl pyrocarbonate [5] (2.4 g, 11 mmol) is added and stirring is continued at room temperature for 30 min. The solution is concentrated in vacuo to about 10 to 15 ml, cooled in an ice-water bath, covered with a layer of ethyl acetate (30 ml) and acidified with a dilute solution of $KHSO_4$ to pH 2–3 (Congo paper). The aqueous phase is extracted with ethyl acetate (15 ml) and the extraction repeated. The ethyl acetate extracts are pooled, washed with water (twice, 30 ml each time), dried over anhydrous Na_2SO_4 and evaporated in vacuo. The residue is recrystallized with a suitable solvent [6].

1. Tarbell DS, Yamamoto Y, Pope BM (1972) Proc Nat Acad Sci USA 69: 730
2. Moroder L, Hallett S, Wünsch E, Keller O, Wersin G (1976): Hoppe Seyler's Z. Physiol Chem 357: 1651
3. The application of *tert*-butyl pyrocarbonate for the introduction of the Boc group could be further improved by the use of a pH-stat (Perseo G, Piani S, de Castiglione R (1983): Int J Peptide Protein Res 21: 227). This, in turn, is an adaptation of the method of Schnabel (Justus Liebigs (1967) Ann Chem 702: 108) in which the reaction between the *tert*-butyl azidoformate and the amino acid was similarly monitored. The single drawback of these methods is that the optimal pH has to be determined for each amino acid. It should be noted that *tert*-butyl azidoformate exploded in several laboratories and is no longer commercially available. It can be prepared, however, from *tert*-butyl carbazate and be used without isolation.
4. In the case of alkali sensitive amino acids, such as asparagine of glutamine it might be advantageous to use Na_2CO_3 (1.06 g, 10 mmol) in water (10 ml) instead of NaOH.
5. Commercially available, designated as di-*tert*-butyl dicarbonate. It melts at 20–22 °C and should be stored in a refrigerator.

6. If crystallization, e.g. from ethyl acetate-hexane fails, the dicyclohexylammonium salt can be prepared by adding dicyclohexylamine (1.8 g, 2.0 ml, 10 mmol) to a solution of the BOC-amino acid in absolute ethanol and diluting the solution with ether.

1.6 The Biphenylylisopropyloxycarbonyl (Bpoc) Group [1, 2]

Bpoc-L-leucine [1, 2]

pyridine

$C_{22}H_{27}NO_4$ (369.5)

A solution of *p*-biphenylyl-dimethyl-carbinol [3] (21.2 g, 100 mmol) in dichloromethane (100 ml) and pyridine (12 ml) is stirred and cooled to $-5\,^{\circ}C$ and phenyl chlorocarbonate [4] (15.2 ml = 18.9 g, 120 mmol) in dichloromethane (50 ml) is added, dropwise, over a period of approx. 30 min. Stirring is continued at 0 °C overnight. During this time most of the heavy precipitate which formed during the addition of the chlorocarbonate dissolves. The mixture is poured onto some cracked ice, dichloromethane (100 ml) is added, the organic phase separated and washed three times with water (200 ml each time) and dried over anhydrous Na_2SO_4. The solvent is removed in vacuo at a bath temperature of 30 °C and the residue is taken up in ether (100 ml). The solution is concentrated in vacuo to about 60 ml then cooled in an ice-water bath for crystallization. The first crop weighs 26 g (78%) and melts at 115–116 °C dec. On concentration of the mother liquor to about 10 ml a second crop, 3.2 g (10%) melting at 114–115 °C is obtained [5].

L-Leucine (13.1 g, 100 mmol) is dissolved in a 2.5 N methanolic solution of benzyltrimethylammonium hydroxide [6] (40 ml, 100 mmol) and the solvent is removed in vacuo. The residue is dissolved in dimethylformamide (30 ml), the solution evaporated to dryness in vacuo and the addition and evaporation of dimethylformamide repeated. The dry salt is redissolved in dimethylformamide (40 ml), warmed to 50 °C and treated with the mixed carbonate (33.3 g, 100 mmol) described in the previous paragraph. The reaction mixture is stirred at 50 °C for 3 hours then cooled in an ice-water bath, diluted with water (200 ml) and ether (200 ml) is added. The aqueous phase is acidified to pH 2–3 with a one M solution of citric acid in water, the layers separated and the aqueous solution reextracted with ether (twice, 100 ml each time). The combined ether extracts are washed with water (twice 100 ml each time), dried over anhydrous Na_2SO_4 and evaporated in vacuo at a bath temperature of

30 °C. The crystalline residue is triturated with ether (15 ml) and petroleum ether (b.p 40–60 °C, 50 ml) and insoluble Bpoc-L-leucine is collected by filtration: 26 g (70%). It melts at 227–230 °C dec. and gives the expected values on elemental analysis.

Bpoc-L-leucine [7]

$CH_3O{-}CO{-}C_6H_4{-}OH + COCl_2 + (CH_3)_2N{-}C_6H_5 \longrightarrow CH_3O{-}CO{-}C_6H_4{-}O{-}CO{-}Cl + C_6H_5{-}N(CH_3)_2{\cdot}HCl$

$C_6H_5{-}C_6H_4{-}C(CH_3)_2{-}OH + Cl{-}CO{-}O{-}C_6H_4{-}CO{-}OCH_3 \xrightarrow{\text{pyridine}} C_6H_5{-}C_6H_4{-}C(CH_3)_2{-}O{-}CO{-}O{-}C_6H_4{-}CO{-}OCH_3$ (+ Cl.HNC_5H_5)

$C_6H_5{-}C_6H_4{-}C(CH_3)_2{-}O{-}CO{-}O{-}C_6H_4{-}CO{-}OCH_3 + H_2N{-}CH(CH_2{-}CH(CH_3)_2){-}COOH \longrightarrow$

$C_6H_5{-}C_6H_4{-}C(CH_3)_2{-}O{-}CO{-}NH{-}CH(CH_2{-}CH(CH_3)_2){-}COOH$

$C_{22}H_{27}NO_4$ (369.5)

A solution of phosgene (25 g, 250 mmol) in dry benzene (100 ml) is stirred at room temperature and methyl 4-hydroxybenzoate (30.4 g, 200 mmol) is added followed by the dropwise addition of *N,N*-dimethylaniline (24.4 g – 25.6 ml, 200 mmol). The temperature of the reaction mixture is maintained at or below 20 °C by cooling and regulating the rate of addition of the base. Stirring is continued at room temperature for three hours. The mixture is filtered and the precipitated dimethylaniline hydrochloride washed with dry benzene. The combined filtrate and washings are extracted, first with N HCl (200 ml) and then twice with water (200 ml each time). The organic layer is dried over $CaCl_2$, and the solvent removed in vacuo. The residue is distilled under reduced pressure and the fraction boiling at 150 °C and 12 mm is collected. On cooling the purified 4-methoxycarbonylphenyl chlorocarbonate solidifies in crystalline form. It weighs 28 g (65%) and melts at 51–52 °C. The i.r. spectrum shows two carbonyl bands, at 1717 and 1765 cm^{-1}.

2-(4-biphenylyl)-2-propanol (21.2 g, 100 mmol) is dissolved in dichloromethane (100 ml) and dry pyridine (8 ml), the solution is cooled to 0–4 °C and stirred while 4-methoxycarbonylphenyl chlorocarbonate (21.4 g, 100 mmol) in dichloromethane (100 ml) is added dropwise. Stirring is continued at about 4 °C for 5 hours. Ice water (about 200 ml) is added, the organic layer washed with N HCl and then several times with water. It is dried over anhydrous Na_2SO_4 and evaporated in vacuo to dryness. The crystalline residue, 2-(4-biphenylyl)-2-propyl 4-methoxycarbonylphenyl carbonate [8] is recrystallized from ethyl acetate-ether and from dichloromethane-ether. The purified mixed carbonate, 30 g (77%) melts, on rapid heating, at 124–126 °C dec.; in the i.r. spectrum the characteristic carbonyl bands appear at 1728 and 1760 cm^{-1}.

L-Leucine (1.3 g, 10 mmol) is dissolved, with warming, in a 40% methanolic solution of Triton B [6] (4.7 ml), the solvent removed in vacuo, the residue dissolved in dimethylformamide (5 ml), evaporated in vacuo, and this process, used for the removal of methanol, repeated. The amino acid salt is redissolved in dimethylformamide (10 ml) and treated with the mixed carbonate described in the previous paragraph (3.9 g, 10 mmol) at 50 °C for 3 hours. The mixture is allowed to cool to room temperature, diluted with ether (125 ml) and *N*-methylmorpholine (0.6 ml) is added followed by ice-water and 1 M citric acid to produce a pH 2–3 in the aqueous phase. The ether layer is washed with ice water until the washes are no more acidic, the ether is removed in vacuo and the residue triturated with ether and petroleum ether. The resulting solid is collected on a filter: 3.35 g (91%). The protected amino acid changes its form at 150 °C and melts with dec. at 224–226 °C; $[\alpha]_D^{22}$ −12.2° (*c* 1, methanol). On elemental analysis correct values are obtained for C, H and N.

1. Sieber P, Iselin B (1968) Helv Chim Acta 51: 622.
2. The abbreviation Bpoc is used for 2-(*p*-biphenylyl)-isopropyloxycarbonyl.
3. Mowry DT, Dazzi J, Renoll M, Shortridge RW (1948): J Amer Chem Soc 70: 1916.
4. Commercially available, usually as phenyl chloroformate.
5. Recrystallization from ether yields analytically pure material melting at 115 to 116 °C dec.
6. Triton B, commercially available in methanolic solution.
7. Schnabel E, Schmidt G, Klauke E (1971): J Liebigs Ann Chem 743: 69.
8. This mixed carbonate is more reactive than the phenyl derivative described in ref. 1 and on p.21. It is also less sensitive to water and less susceptible to thermal decomposition (cf. Sieber P, Iselin B (1969): Helv Chem] Acta 52: 1525). 4-Methoxycarbonylphenyl-2-(4-biphenylyl)-2-propyl carbonate can be stored at room temperature for at least two months.

1.7 The 9-Fluorenylmethyloxycarbonyl (Fmoc) Group [1]

9-Fluorenylmethyl Chlorocarbonate [1]

H CH_2OH + $COCl_2$ → H $CH_2O{-}CO{-}Cl$ + HCl

$C_{15}H_{11}O_2Cl$ (258.8)

A solution of phosgene (10.9 g, 110 mmol) in dichloromethane (110 ml) is stirred and cooled in an ice-water bath. 9-Fluorenylmethanol [2] (19.6 g,

100 mmol) is added in small portions and stirring is continued for about one hour. The mixture is kept in the ice-water bath for an additional four hours when the solvent and excess phosgene are removed in vacuo. The residue, an oil, slowly solidifies. The chlorocarbonate (24.6 g, 95%) melts at 61.5–63 °C. Two recrystallizations from ether afford colorless crystals (22.3 g, 86%) with unchanged melting point. On elemental analysis correct values are obtained.

9-Flurenyl-methyloxy-carbonyl-L-tryptophan [1]

H CH$_2$O–CO–Cl + H$_2$N–CH(CH$_2$-indolyl)–COONa → (1. Na$_2$CO$_3$; 2. HCl) → H CH$_2$O–CO–NH–CH(CH$_2$-indolyl)–COOH

$C_{26}H_{22}N_2O_4$ (426.5)

L-Tryptophan (2.04 g, 10 mmol) is dissolved in a 10% solution of Na_2CO_3 in water (26.5 ml, 25 mmol). Dioxane (15 ml) is added and the mixture is stirred in an ice-water bath. 9-Fluorenylmethyl chlorocarbonate [3] (2.6 g, 10 mmol) is added in small portions and stirring is continued at ice-water bath temperature for four hours and then at room temperature for eight hours. The reaction mixture is poured into water (600 ml) and extracted with ether (200 ml in two portions). The aqueous solution is cooled in an ice-water bath and acidified under vigorous stirring with concentrated hydrochloric acid to Congo. The mixture is stored in the refrigerator overnight, filtered and the solid material thoroughly washed with water. The dry product (4.0 g, 94%) melts at 182–185 °C. Recrystallization from nitromethane and a second recrystallization from chloroform-hexane yields analytically pure material (3.9 g, 91%) which melts at 185–185 °C dec.; $[\alpha]_D^{24}$ +6.4 (*c* 1, ethyl acetate) [4].

1. Carpino LA, Han GY (1972) J Org Chem 37: 3404
2. Brown WG, Bluestein BA (1953): J Amer Chem Soc 65: 1082

3. The chlorocarbonate is commercially available.
4. Preparation of Fmoc-amino acids is accompanied by the formation of small amounts of Fmoc-dipeptides, obviously via mixed anhydrides generated from a reaction between the Fmoc-amino acid already formed and still unreacted 9-fluorenylmethyl chloroformate. This side reaction occurs during the introduction of other blocking groups as well, if, instead of alkali, bicarbonate is used as acid binding agent.

1.8 *o*-Nitrophenylsulfenylamino Acids [1, 2]

Dicyclohexyl-ammonium salts

o-NO$_2$-C$_6$H$_4$–S–Cl + H$_2$N–CHR–COONa → (1. NaOH; 2. H$_2$SO$_4$) → o-NO$_2$-C$_6$H$_4$–S–NH–CHR–COOH → ((C$_6$H$_{11}$)$_2$NH) → o-NO$_2$-C$_6$H$_4$–S–NH–CHR–COO$^-$·H$_2\overset{+}{N}$(C$_6$H$_{11}$)$_2$

The amino acid (20 mmol) is dissolved in a mixture of 2 N NaOH (10 ml) and dioxane (25 ml). The solution is stirred and *o*-Nitrophenylsulfenyl chloride (4.2 g, 22 mmol) is added in ten approximately equal portions with the simultaneous dropwise addition of 2 N NaOH (12 ml). About 15 minutes are needed for the addition of the reactants. The mixture is diluted with water (200 ml), filtered and acidified to Congo with one normal sulfuric acid. The protected amino acids usually separate as syrups which crystallize on scratching and cooling. The product is collected on a filter, washed with water, dissolved in ethyl acetate and precipitated with hexane. Alternatively, the acidified reaction mixture is extracted with a 1:1 mixture of ethyl acetate and ether, the organic phase washed with water until the washes are neutral to Congo paper. The solution is dried over anhydrous Na_2SO_4 and treated with dicyclohexylamine (3.64 g, 20 mmol). The dicyclohexylammonium salt which separates is collected, washed with ethyl acetate and with ether and dried in air [3].

1. Zervas L, Borovas D, Gazis E (1963) J Amer Chem Soc 85: 3660
2. This is the designation commonly used in the literature; *o*-nitrobenzenesulfenyl is, however, more appropriate.
3. Salts of *o*-nitrophenylsulfenyl amino acids are more stable than the free acids. The free acids can be obtained from their salts by dissolving or suspending them in a mixture of water and ethyl acetate and adding a 2% solution of $KHSO_4$ until the aqueous layer is acidic to Congo. The organic layer is washed with water and the protected amino acid is obtained by evaporation of the solvent in vacuo.

1.9 2-Trimethylsilylethyloxycarbonylamino Acids [1]

4-Nitrophenyl-2-Trimethylsilylethyl Carbonate [2]

$$(CH_3)_3SiCH_2{-}CH_2OH + COCl_2 \longrightarrow (CH_3)_3SiCH_2{-}CH_2O{-}CO{-}Cl + HCl$$

$$(CH_3)_3SiCH_2{-}CH_2O{-}CO{-}Cl + HO{-}C_6H_4{-}NO_2 + N(C_2H_5)_3 \longrightarrow$$

$$(CH_3)_3SiCH_2{-}CH_2O{-}\overset{O}{\overset{\|}{C}}{-}O{-}C_6H_4{-}NO_2 + (C_2H_5)_3NH \cdot Cl$$

$C_{12}H_{17}NO_5Si$ (283.4)

A solution of phosgene [3] (123 g, 1.25 mol) in dry toluene [4] (600 ml) is cooled to $-25\,^\circ C$ and 2-trimethylsilylethanol [5] (118.3 g = 143 ml, one mol) in dry toluene [4] (100 ml) is added dropwise, with stirring, over a period of approx. 30 min. During the addition of the alcohol the temperature of the reaction mixture is kept at about $0\,^\circ C$. The mixture is stirred at $-5\,^\circ C$ for one and a half hours, then dry nitrogen is bubbled through the solution [3] at room temperature for two days. The mixture is cooled to about $-15\,^\circ C$ and a solution of *p*-nitrophenol (146 g, 1.08 mol) and triethylamine (150 ml = 109 g, 1.08 mol) in ethyl acetate (150 ml) is slowly added. The temperature of the mixture is maintained below $-3\,°C$ by the rate of addition of the reactants. Stirring is continued at room temperature for one and a half hours and the precipitate (triethylammonium chloride) is removed by filtration. The solution is washed with an ice cold (about 3%) solution of $KHSO_4$ in water until the washes are acidic, then with an ice cold, saturated solution of $NaHCO_3$ and finally with ice cold, saturated solution of NaCl. The organic layer is dried over anhydrous Na_2SO_4 and evaporated in vacuo to dryness. The crystalline product is dissolved in petroleum ether (b.p. 40–60 °C) the solution is filtered from some insoluble material and concentrated under reduced pressure. During evaporation the mixed carbonate separates in crystalline form. The crystals are collected, washed with cold (−40 °C) petroleum ether and dried. The product, 165 g (58%) melts at 34–36 °C. The i.r. spectrum shows the characteristic active ester carbonyl band at 1765 cm^{-1}.

2-Trimethylsilylethyloxycarbonyl-L-proline [2]

$(CH_3)_3SiCH_2{-}CH_2O{-}C(=O){-}O{-}C_6H_4{-}NO_2$ + HN—CH—COONa (pyrrolidine ring) $\xrightarrow{\text{1. NaOH; 2. }KHSO_4}$

$(CH_3)_3SiCH_2{-}CH_2O{-}CO{-}N{-}CHCOOH$ (pyrrolidine ring) $\xrightarrow{(C_6H_{11})_2NH}$

$(CH_3)_3SiCH_2{-}CH_2{-}O{-}CO{-}N{-}CH{-}COOH\cdot HN(C_6H_{11})_2$ (pyrrolidine ring)

$C_{23}H_{44}N_2O_4Si$ (440.7)

L-Proline (1.15 g, 10 mol) is dissolved in N NaOH, a solution of 4-nitrophenyl 2-trimethylsilylethyl carbonate (2.84 g, 10 mmol) in *tert*-butanol [6] (20 ml) is added and the mixture is stirred at room temperature overnight. Most of the organic solvent is removed in vacuo and the solution diluted with water (about 40 ml). Ether [7] (about 50 ml) is added and enough 2% $KHSO_4$ to lower the pH to about 2 to 3. The two layers are separated and the aqueous layer reextracted with ether [7] (50 ml). The combined extracts are washed with water (50 ml), dried over anhydrous Na_2SO_4 and concentrated in vacuo to about 50 ml. A solution of dicyclohexylamine (1.81 g = 2.0 ml, 10 mmol) in ether (20 ml) is added and the crystals which form are collected on a filter, washed with ether (25 ml) and dried in vacuo. The dicyclohexylammonium salt (3.1 g, 70%) melts at 145–148 °C; $[\alpha]_D^{20}$ −28.3° (*c* 1, methanol).

1. Carpino LA, Tsao JH, Ringsdorf H, Fell E, Hettrich G: J Chem Soc Chem Commun 1978: 358
2. Wünsch E, Moroder L, Keller O (1981): Hoppe Seyler's Z Physiol Chem 362: 1289
3. The preparation of the mixed carbonate must be carried out in a well ventilated hood.
4. Toluene can be dried by distillation at atmospheric pressure: water is removed with the forerun.
5. Commercially available.
6. Dioxan can be used instead of *tert*-butanol.
7. Ethyl acetate is equally suitable.

1.10 Maleoylamino Acids and Maleoyl-Peptides [1]

N-Methoxycarbonylmaleimide

$$\text{Maleimide} + Cl{-}CO{-}OCH_3 + CH_3{-}N(\text{morpholine}) \longrightarrow \text{Maleimide-}N{-}CO{-}OCH_3 + CH_3{-}N(\text{morpholine})\cdot HCl$$

$C_6H_5NO_4$ (155.1)

Maleimide (0.97 g, 10 mmol) and N-methylmorpholine (1.01 g = 1.10 ml; 10 mmol) are dissolved in ethyl acetate (50 ml), the solution stirred and cooled in an ice-water bath and methyl chlorocarbonate [2] (0.95 g = 0.78 ml; 10 mmol) is added. Half an hour later the precipitate (N-methylmorpholine hydrochloride) is removed by filtration and washed with ethyl acetate. The combined filtrate and washings are extracted three times with water (50 ml each time), dried over anhydrous Na_2SO_4 and evaporated in vacuo to dryness. The product is recrystallized from ethyl acetate-diisopropyl ether. It weighs 9.9 g (64%) and melts at 61–63 °C. It is homogeneous on thin layer plates of silica gel; R_f 0.67 in chloroform-methanol (1:1); R_f 0.45 in chloroform-acetic acid (19:1) [3]. In the NMR spectrum ($CDCl_3$, Me_4Si) the singlet of the methyl protons appears at 3.93 ppm, the vinyl protons, also a singlet, at 7.0 ppm. Correct analytical values are obtained for C, H and N.

Maleoyl-glycine (Maleimidoacetic Acid)

$$\text{Maleimide-}N{-}CO{-}OCH_3 + H_2N{-}CH_2{-}COONa \longrightarrow \left[\begin{array}{l} CH{-}CO{-}NH{-}CO{-}OCH_3 \\ \| \\ CH{-}CO{-}NH{-}CH_2{-}COONa \end{array}\right] \longrightarrow$$

$$\text{Maleimide-}N{-}CH_2COONa + H_2N{-}CO{-}OCH_3$$

$$\downarrow H_2SO_4$$

$$\text{Maleimide-}N{-}CH_2COOH$$

$C_6H_5NO_4$ (155.1)

Glycine (0.75 g, 10 mmol) is dissolved in a saturated aqueous solution of $NaHCO_3$ (50 ml), the mixture is cooled in an ice-water bath and treated, under vigorous stirring, with finely powdered N-methoxycarbonylmaleimide (1.55 g, 10 mmol). Ten minutes later the solution is diluted with water (200 ml) [4] and stirred at room temperature for 30 to 40 minutes. The pH is adjusted to 6–7 with sulfuric acid, the solution is evaporated in vacuo to about 60 ml and acidified to pH 1–2 with 2 N sulfuric acid and extracted with ethyl acetate. The extracts are washed with water, dried over anhydrous Na_2SO_4 and evaporated in vacuo to dryness. The residue is dissolved in $CHCl_3$ containing 5% acetic acid (10–20 ml), filtered through a column of silica gel (40 g) and eluted with the same solvent [5]. The solvents are removed in vacuo and the residual acetic acid eliminated by evaporation with water. The crude maleoyl-glycine is recrystallized from ether-petroleum ether. The purified product (1.09 g, 70%) melts at 105–106 °C; R_f-s in the two systems described above are 0.40 and 0.12 respectively. On elemental analysis the expected values are found [6].

Maleoyl Peptides

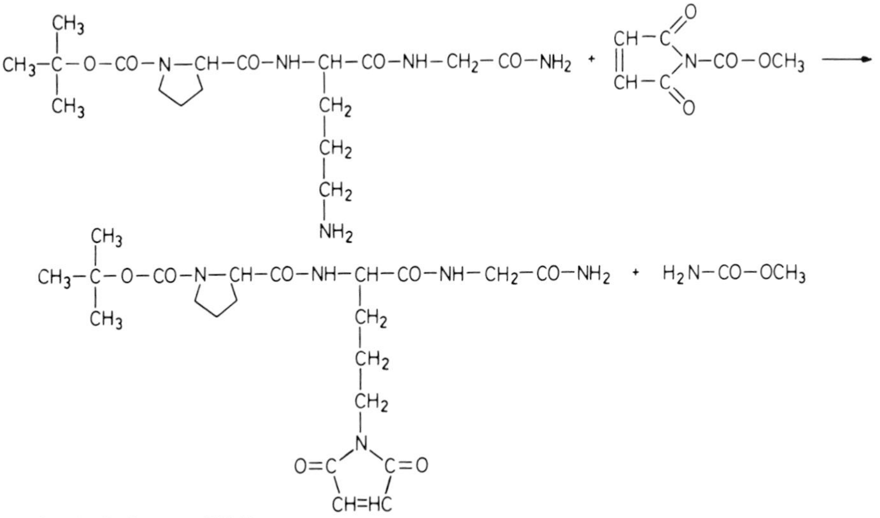

$C_{21}H_{31}N_5O_7$ (465.5)

The partially protected tripeptide amide Boc-Pro-Orn-Gly-NH_2 (385 mg, 1 mmol) is dissolved in saturated aqueous $NaHCO_3$ (5 ml). The solution is stirred and cooled in an ice-water bath and treated with *N*-methoxycarbonyl-maleimide (310 mg, 2 mmol). After 10 minutes at about 0 °C water (25 ml) is added and stirring is continued at room temperature for about 15 minutes. The solution is extracted four times with 4:1 mixture of ethyl acetate-n-butanol (30 ml each time). The combined extracts are washed twice with water (50 ml each time), dried over anhydrous Na_2SO_4 and evaporated in vacuo to about 5 ml. The product is precipitated by the addition of ether. The fully blocked peptide (290 mg, 63%) melts at 209–210 °C, dec.; $[\alpha]_D -68°$ (*c* 1, trifluoroethanol); R_f-s are 0.64 and 0.02 in the systems mentioned under 1. The proton NMR spectrum (in Me_2SO-d_6) shows the characteristic maleimide protons as a two-proton singlet at 6.98 ppm. On elemental analysis the expected values are obtained [7].

1. Keller O, Rudinger J (1975) Helv Chim Acta 58: 531
2. Commercially available, usually under the name of methyl chloroformate.

3. Maleimides can be detected by spraying the thin layer plates with a 0.1% solution of 5,5'-dithio-bis-2-nitrobenzoic acid in 1:1 ethanol-Tris-HCl buffer (pH 8.2) and then with a 2% solution of sodium 2-mercaptoethanesulfonate in 80% ethanol until the background is bright yellow. Maleimide derivatives appear as white spots.
4. In the preparation of the maleoyl derivative of phenylalanine, dioxane or tetrahydrofuran (100 ml) is used for dilution and the mixture is kept at 40 °C rather than at room temperature for 30–40 minutes. In the maleoylation of the ε-amino group of N^{α}-benzyloxycarbonyl-L-lysine dioxane or tetrahydrofuran are used for dilution, but the following stirring is carried out at room temperature.
5. Filtration through silica-gel removes the maleamic acid derivative produced by incomplete ring closure or by partial hydrolysis of the product.
6. Some maleoylamino acids, which do not crystallize, are treated with a slight excess of dicyclohexylamine in ether or acetone. This usually yields a crystalline dicyclohexylammonium salt.
7. The maleoyl group is removed by hydrazinolysis in dilute $NaHCO_3$ solution at 40 °C. For 1 mmol of the maleoyl peptide 2 mmoles of hydrazine are needed. Alternatively, the maleoyl group can be cleaved by ring opening through alkaline hydrolysis (5% Na_2CO_3 solution) at room temperature for about one hour followed by hydrolysis of the maleamic acid thus formed in dioxane-water containing acetic acid in one molar concentration. The hydrolysis with acid requires about 20 hours at 40 °C.

1.11 Triphenylmethyl-amino Acids (Trityl-amino Acids) [1]

Trityl-L-leucine [2]

H_3C–CH(CH_3)–CH_2–CH(NH_2)–COOH → 1. $(CH_3)_3SiCl$; 2. $(C_6H_5)_3CCl$ → $(C_6H_5)_3C$–NH–CH(CH_2–CH(CH_3)$_2$)–COOH

$C_{25}H_{27}NO_2$ (373.5)

Trimethylchlorosilane (1.1 g, 1.27 ml, 10 mmol is added to a stirred suspension of L-leucine (1.31 g, 10 mmol) in chloroform (115 ml) and acetonitrile (3 ml) and the mixture is heated under a reflux condenser for two hours. It is cooled to room temperature and triethylamine (2.02 g, 2.8 ml, 20 mmol) is added slowly, at a rate that maintains gentle reflux, followed by a solution of triphenylchloromethane (triphenylmethyl chloride, trityl chloride, 2.8 g, 10 mmol) in chloroform (10 ml) and the reaction mixture is kept at room temperature for one hour. After the addition of methanol (50 ml) the solvents are evaporated in vacuo. The residue is partitioned between ether (50 ml) and an ice-cold 5% solution (50 ml) of citric acid in water. The organic phase is washed with N NaOH (twice, 20 ml each time) and then with water (twice, 20 ml each time). The combined aqueous layers are washed with ether (20 ml), cooled in an ice-water bath and neutralized with glacial acetic acid. The precipitate, which forms, is extracted into ether (60 ml in two equal portions), the extract washed twice with water, dried over anhydrous magnesium sulfate and evaporated under reduced pressure. The residue, a foam (3.5 g), can be used for coupling or for the preparation of active esters. For storage it is converted to the crystalline diethylammonium salt melting at 152–154 °C; $[\alpha]_D^{25}$ +3.5° (*c* 5, MeOH).

N-Trityl Amino Acids [3]

$$H_2N-CHR-COOH \xrightarrow[N(C_2H_5)_3]{2\,(C_6H_5)_3CBr} (C_6H_5)_3C-NH-CHR-C(=O)-O-C(C_6H_5)_3$$

$$\xrightarrow{CH_3OH} (C_6H_5)_3C-NH-CHR-COOH \quad + \quad H_3CO-C(C_6H_5)_3$$

In a dry round-bottom flask, the suspension of the finely-powdered amino acid (10 mmol) in a solution of triphenylbromomethane (triphenylmethyl bromide, tritylbromide, 7.11 g, 22 mmol) in a 2:1 mixture of chloroform and dimethylformamide (50 ml), is vigorously stirred at room temperature until a clear solution is obtained. This usually requires from one half to one hour. A solution of triethylamine (4.04 g, 6.6 ml, 40 mmol) in a 2:1 mixture of chloroform and dimethylformamide (40 ml) is added dropwise, for about 20 minutes, and stirring is continued for a further 30 minutes. After addition of methanol (50 ml) the mixture is heated to 50 °C for a period ranging from 20 minutes to 2.5 hours [4] and then the solvent is evaporated in vacuo. The residue is taken up in ether (100 ml) and the solution washed with three portions (each 50 ml) of a 10% solution of citric acid in water, with water (three times, 50 ml each time) and dried over anhydrous sodium sulfate. A solution of diethylamine (0.73 g, 1.05 ml, 10 mmol) or dicyclohexylamine (1.82 g, 2.0 ml, 10 mmol) in ether (10 ml) is added dropwise. The precipitated salts are filtered, washed several times with cold ether and dried. The products are chromatographically homogeneous and have the expected NMR spectra. Yields range from 80 to 86%.

1. Helferich B, Moog L, Junger A (1925) Ber dtsch Chem Ges 58: 872
2. Barlos K, Papaioannou D, Theodoropoulos D (1982): J Org Chem 47: 1324
3. Mutter M, Hersperger R (1989): Synthesis, 198
4. In the preparation of trityl-glycine, methanolysis required only 20 minutes but 1.5 hours in the case of other amino acids and 2.5 hours for amino acids with bulky side chains.

2 Blocking of the α-Carboxyl Group

2.1 Methyl Esters [1, 2]

$$H_2N{-}CHR{-}COOH + (CH_3O)_2C(CH_3)_2 + HCl(aq) \longrightarrow HCl{\cdot}H_2N{-}CHR{-}CO{-}OCH_3 + CH_3COCH_3$$

The amino acid (10 mmol) is suspended in 2,2-dimethoxypropane [3] (100 to 150 ml) and concentrated hydrochloric acid (10 ml) is added. The mixture is allowed to stand [4] at room temperature overnight. The volatile reactants are removed in vacuo at a bath temperature not exceeding 60 °C, the residue dissolved in a minimum amount of dry methanol and the solution diluted with dry ether (250 ml). The crystalline methyl ester hydrochloride is collected on a filter, washed with ether and dried in vacuo over NaOH pellets. Recrystallization from methanol-ether affords the analytically pure ester hydrochloride [5]; yields range from 80 to 95%. The values of specific rotations agree with those reported in the literature.

Consideration of the various methods

1. Rachele JR (1963) J Org Chem 28: 3898
2. The classical method of esterification of amino acids with methanol or ethanol (Curtius T (1883) Ber dtsch Chem Ges 16: 753; Fischer E (1901) Ber dtsch Chem Ges 34: 433) requires dry HCl gas. Esterification with the aid of thionyl chloride (Brenner M, Huber W (1953) Helv Chim Acta 36: 1109) yields products which must be purified, e.g., by distillation of the amino acid esters. Such considerations led to the selection of the simple procedure of Rachele (ref. 1) for this volume. Esterification of amino acids with methanol in the presence of excess *p*-toluenesulfonic acid is also possible (Bodanszky M, Bodanszky A, unpublished). Thus a suspension of L-leucine (1.31 g, 10 mmol) in methanol (50 ml) was treated with *p*-toluenesulfonic acid monohydrate (3.8 g, 20 mmol), and the resulting solution boiled under reflux for 24 hours. Evaporation to dryness and trituration of the residue with ether (100 ml) followed by washing with ether yielded L-leucine methyl ester *p*-toluenesulfonate (3.0 g, 95%) melting at 172–174 °C; $[\alpha]_D^{22}$ +7.8° (*c* 5, H_2O); +11.6° (*c* 6.9, methanol). Recrystallization from hot acetone raised the m.p. to 173–175 °C. The same compound obtained through the reaction of the amino acid with methyl sulfite in the presence of *p*-toluenesulfonic acid (Theobald JM, Williams MW, Young GT (1927) J Chem Soc 1963) melted at 175.5–176 °C; $[\alpha]_D^{21}$ +11.6° (*c* 6.9 methanol).
3. Acetone dimethyl ketal. The commercially available material should be redistilled (b.p. 79–81 °C).
4. In the case of lysine (monohydrochloride) methanol (30–40 ml) is added to enhance its solubility and the mixture is heated to reflux for 2 hours prior to storage at room temperature

overnight. The same modification, but with a reflux period of 5 hours, is needed in the preparation of glutamic acid dimethyl ester (hydrochloride) from glutamic acid.
5. L-Lysine methyl ester dihydrochloride prepared by this procedure crystallized with one mole of methanol. The product melted at 60–75 °C, resolidified to melt finally at 199–200 °C. Recrystallization from a small volume of water by the addition of acetone and 2,2-dimethoxyethane afforded unsolvated methyl ester dihydrochloride.

2.2 Ethyl Esters

2.2.1 Esterification with the Help of Gaseous HCl [1]

L-Tyrosine Ethyl Ester [2]

$$HO\text{-}C_6H_4\text{-}CH_2\text{-}CH(NH_2)\text{-}COOH + CH_3CH_2OH + HCl \longrightarrow Cl\cdot H_3N\text{-}CH(CH_2\text{-}C_6H_4\text{-}OH)\text{-}CO\text{-}OC_2H_5 \xrightarrow{NaOH} H_2N\text{-}CH(CH_2\text{-}C_6H_4\text{-}OH)\text{-}CO\text{-}OC_2H_5$$

$C_{11}H_{15}NO_3$ (209.3)

Dry HCl gas is introduced into a suspension of L-tyrosine (18.1 g, 100 mmol) in absolute ethanol (125 ml) without cooling until a clear solution forms. More absolute ethanol (250 ml) is added and the reaction mixture is boiled under reflux for about 6 hours. The solution is cooled to room temperature and evaporated to dryness in vacuo. The residue is dissolved in absolute ethanol (100 ml) and reevaporated to dryness. The hydrochloride salt is dissolved in water (100 ml), the solution is cooled in an ice-water bath and neutralized with 5 N NaOH (about 20 ml). The ethyl ester separates in crystalline form. It is collected on a filter, washed with cold water (about 20 ml in several portions) and dried in vacuo over P_2O_5. After recrystallization from ethyl acetate the purified L-tyrosine ethyl ester (about 17 g, 81%) melts at 108–109 °C; $[\alpha]_D^{20}$ +20.4° (*c* 4.9, abs. ethanol) [3].

1. Curtius T (1883) Ber dtsch Chem Ges 16: 753
2. Fischer E (1901) Ber dtsch Chem Ges 34: 433; (1908) 41: 850
3. Generally amino acid ester hydrochlorides are the target compounds that can be purified by recrystallization from methanol-ether. Isolation of the ester itself is possible in the case of tyrosine because of the formation of a phenolate salt.

2.2.2 Esterification Catalyzed by *p*-Toluenesulfonic Acid [1]

L-Methionine Ethyl Ester *p*-Toluenesulfonate [2, 3]

$$CH_3\text{-}S\text{-}CH_2\text{-}CH_2\text{-}CH(NH_2)\text{-}COOH + C_2H_5OH + CH_3\text{-}C_6H_4\text{-}SO_3H \xrightarrow{(-H_2O)} CH_3\text{-}C_6H_4\text{-}SO_3H \cdot H_2N\text{-}CH(CH_2\text{-}CH_2\text{-}S\text{-}CH_3)\text{-}CO\text{-}OC_2H_5$$

$C_{14}H_{23}NO_5S_2$ (349.4)

To a suspension of L-methionine (1.49 g, 10 mmol) in absolute ethanol (50 ml) *p*-toluenesulfonic acid (monohydrate, 3.8 g 20 mmol) is added and the resulting solution heated under reflux for 24 hours. The alcohol is removed in the vacuum of a water aspirator and the residue triturated with dry ether, free from peroxides (100 ml). The crystalline material is thoroughly washed on a filter with ether (100 ml) and dried in vacuo over P_2O_5. It weighs 3.44 g (98%) and melts at 124–126 °C; $[\alpha]_D^{22} + 13°$ (*c* 3.3, 95% ethanol).

1. The procedure described here (Bodanszky M, Bodanszky A, unpublished) is an adaptation of the method used for the preparation of benzyl esters; cf. Miller HK, Waelsch H (1952) J Amer Chem Soc 74: 1902
2. Methods of esterification in which alkylating agents are applied are not suitable for the preparation of esters of methionine, tyrosine or tryptophan.
3. By the same process L-leucine ethyl ester *p*-toluenesulfonate (m.p. 156–158 °C; $[\alpha]_D^{22} + 10.3°$ (*c* 3.5, 95% ethanol) was obtained in 90% yield and the *p*-toluenesulfonate salt of ethyl L-tryptophanate, melting at 138–143°, in 96% yield. Recrystallization of this material from abs. ethanol-ether afforded the pure ester-salt melting at 142–143 °C; $[\alpha]_D^{22} + 18.1°$ (*c* 3.2, 95% ethanol).

2.3 Benzyl Esters

Glycine Benzyl Ester *p*-Toluenesulfonate [1]

$$H_2N\text{-}CH_2\text{-}COOH + HOCH_2\text{-}C_6H_5 + CH_3\text{-}C_6H_4\text{-}SO_3H \xrightarrow{-H_2O} CH_3\text{-}C_6H_4\text{-}SO_3H \cdot H_2N\text{-}CH_2\text{-}CO\text{-}OCH_2\text{-}C_6H_5$$

$C_{16}H_{19}NO_5S$ (337.4)

Glycine (18.8 g, 250 mmol) and *p*-toluenesulfonic acid (monohydrate, 48.5 g, 255 mmol) are added to a mixture of freshly distilled benzyl alcohol (100 ml) and benzene [2] (50 ml) in a 500 ml round-bottom flask. The mixture is heated to reflux and the water formed in the reaction trapped in a Dean-Stark receiver [3]. When no more water appears in the distillate the mixture is allowed to cool to room temperature, diluted with ether (500 ml) and cooled in an ice water bath for two hours. The crystalline *p*-toluenesulfonate of glycine benzyl ester is collected on a filter, washed with ether (200 ml) and dried in air. After recrystallization from methanol-ether the salt (30.4 g, 90%) melts at 132–134 °C.

tert-Butyloxycarbonyl-L-asparagine Benzyl Ester [4]

$$CH_3-C(CH_3)_2-O-CO-NH-CH(CH_2CONH_2)-COOCs + BrCH_2-C_6H_5 \longrightarrow CH_3-C(CH_3)_2-O-CO-NH-CH(CH_2CONH_2)-CO-OCH_2-C_6H_5 + CsBr$$

$C_{16}H_{22}N_2O_5$ (322.4)

Water (40 ml) is added to a solution of *tert*-butyloxycarbonyl-L-asparagine (23.2 g, 100 mmol) in methanol (400 ml). The solution is neutralized [5] with a 20% solution of Cs_2CO_3 in water and then evaporated in vacuo to dryness. Dimethylformamide (250 ml) is added and removed in vacuo at a bath temperature of 45 °C. Addition and evaporation of dimethylformamide is repeated and the remaining solid cesium salt of *tert*-butyloxycarbonyl-L-asparagine is treated with dimethylformamide (250 ml) and benzyl bromide [6] (18.8 g = 13.1 ml, 110 mmol). The mixture is stirred at room temperature for 6 hours, evaporated to dryness in vacuo and the residue triturated with water (1 liter). The solid which separates is transferred into ethyl acetate (300 ml), the organic phase washed with water (150 ml), dried over anhydrous Na_2SO_4 and evaporated to dryness in vacuo. The crude ester is recrystallized from ethyl acetate-hexane. The recovered, analytically pure benzyl ester (about 29 g, 90%) melts at 120–122 °C; $[\alpha]_D^{25}$ −17.3 (*c* 1, DMF).

1. Zervas L, Winitz M, Greenstein JP (1955) J Org Chem 22: 1515
2. Benzene is quite toxic. It can be replaced by toluene. The operations should be carried out under a well ventilated hood.
3. Named in some catalogues as a Dean-Stark trap.
   If no such trap is available the mixture is distilled (at atmospheric pressure) and water is repeatedly separated from the distillate while the benzene (or toluene) is returned to the flask. About 9 ml water should be collected.
4. Wang SS, Gisin BF, Winter DP, Makofske R, Kulesha ID, Tzougraki C, Meienhofer J (1977) J Org Chem 42: 1286. The same procedure can also be applied for the preparation of methyl, trityl, α-methylphenacyl and hydroxyphthalimide esters.
5. A moist indicator paper should be used in the titration.
6. Benzyl bromide is one of the worst lachrymators. The esterification must be carried out in a well ventilated hood.

L-Glutamic Acid α-Benzyl Ester [1] (α-Benzyl-L-glutamate)

$$HOOC-CH_2-CH_2-CH(NH_2)-COOH + HOCH_2-C_6H_5 \xrightarrow{HCl} C_6H_5CH_2O-CO-CH_2-CH_2-CH(NH_2\cdot HCl)-CO-OCH_2-C_6H_5$$

$$\xrightarrow{1.\ HI;\ 2.\ N(C_4H_9)_3} HOOC-CH_2-CH_2-CH(NH_2)-CO-OCH_2-C_6H_5$$

$C_{12}H_{15}NO_4$ (237.2)

A suspension of L-glutamic acid (14.7 g, 100 mmol) in freshly distilled benzyl alcohol (220 ml) is warmed to 55 °C and dry HCl is passed into the vigorously stirred mixture, without cooling, for about one hour. Benzene [2] (110 ml) is added and then removed by distillation in vacuo [3]. The mixture is kept under reduced pressure at a bath temperature of about 85 °C for one hour. Dry HCl is passed into the mixture a second time, again for about one hour. The reaction mixture is cooled to room temperature and some unchanged glutamic acid (about 3 g) is removed by filtration. Once more benzene is added (about 100 ml) and removed in vacuo. Dry HCl is passed into the solution a third time followed by the removal of about half of the benzyl alcohol in vacuo. The diester hydrochloride is precipitated by cooling the mixture to room temperature and dilution with ether (about 700 ml). The product is collected by filtration, washed with ether and recrystallized from methanol-ether to yield analytically pure L-glutamic acid dibenzyl ester hydrochloride (about 22 g, 60%) melting at 100–102 °C; $[\alpha]_D^{22} + 9.4°$ (*c* 1.5, 0.1 N HCl).

A 50 mmol aliquot (18.2 g) of the diester hydrochloride is dissolved in acetic acid (200 ml). Constant boiling HI (d = 1.7, 20 ml, 24 mmol) is added and the temperature of the mixture is raised to 50 °C and maintained there for about six hours. The solvent is removed in vacuo and the residual oil is dried by the addition of benzene (100 ml) and distillation of the benzene in vacuo. Dilution with and removal of the benzene is repeated three more times. The remaining dark syrup is diluted with precooled (−10 °C) 95% ethanol (110 ml) containing tri-*n*-butylamine (13 ml = 10.0 g, 54 mmol). More tri-*n*-butylamine (about 5 to 7 ml) is added, enough to produce a neutral reaction on a moist indicator paper. Glutamic acid α-benzyl ester begins to separate. Crystallization is completed by storing the mixture in the refrigerator overnight. The crystals are collected on a filter, thoroughly washed with liberal amounts of absolute ethanol and then with ether. The air-dry product (10–11 g) is dissolved, at room temperature, in 3 N HCl (21 ml, 63 meq), treated with activated charcoal (about 0.5 g) filtered, diluted with an equal volume of absolute ethanol and neutralized with tri-*n*-butylamine. The mixture is kept in the refrigerator for several hours. The crystals are collected and washed with cold absolute ethanol. The α-monobenzyl ester of L-glutamic acid (about 8 g, 67%) melts at 147–148 °C; $[\alpha]_D^{25}$ +12.0° (*c* 3, 0.1 N HCl).

1. Sachs H, Brand E (1953) J Amer Chem Soc 75: 4610
2. Benzene is quite toxic and should be handled with care. The operations described above should be carried out in a well ventilated hood, also on account of the HCl gas used.
3. Water formed in the reaction is eliminated from the mixture in this process and the equilibrium thus shifted toward complete esterification.

2.3.1 Polymeric Benzyl Esters [1] (Anchoring of an Amino Acid to a Polymeric Support by Transesterification [2])

Hydroxymethyl-Polymer [3–5]

CH_3COOK + $ClCH_2$–C₆H₄–(polymer) ⟶ CH_3–CO–O–CH_2–C₆H₄–(polymer) $\xrightarrow{NaOH}$ HO–CH_2–C₆H₄–(polymer)

Chloromethylated styrene-divinylbenzene copolymer [6] (1 or 2% crosslinked [7], about 2 meq. per g) (20 g) is suspended in benzyl alcohol (150 ml). Anhydrous potassium acetate (10 g, about 100 mmol) is added and the suspension is heated to 80 °C and kept at this temperature for about 24 hours. A small sample, filtered and thoroughly washed with ethanol, is tested (Beilstein) for the presence of chlorine [8]. If the test is positive heating is continued until all the chlorine has been displaced by acetate. The resin is collected on a filter, washed with water (twice, 100 ml each time), with dimethylformamide (twice, 50 ml each time) and with methanol (twice, 100 ml each time) [9]. A pellet of the acetoxymethyl resin shows in the i.r. spectrum a strong ester carbonyl band at 1725 cm^{-1}.

The acetoxymethyl resin is suspended in a 2 N solution of NaOH in ethanol (150 ml) and refluxed in a nitrogen atmosphere until no more ester can be detected in the i.r. spectrum of a sample [10]. The resin is washed [9] on a filter with water until the washings are neutral, then with dimethylformamide (twice, 50 ml each time), with 95% ethanol (twice 100 ml) and methanol (twice 100 ml). The resin is dried in air and finally in vacuo. A sample dried over P_2O_5 at 60 °C overnight shows strong OH stretching absorption at 3500 cm^{-1} and the absence of the CO-band at 1725 cm^{-1}.

Transesterification [2, 4, 11] *tert*-Butyloxycarbonyl-glycyl Resin [2]

$(CH_3)_3C$–O–CO–NH–CH_2–CO–O–C₆H₄–NO_2 + HO–CH_2–C₆H₄–(polymer) $\xrightarrow[\text{toluene}]{\text{imidazole}}$ $(CH_3)_3C$–O–CO–NH–CH_2–CO–O–CH_2–C₆H₄–(polymer) + HO–C₆H₄–NO_2

A sample of the hydroxymethylpolymer (cf. above, 3.0 g) is swelled in toluene [12] (25 ml). *tert*-Butyloxycarbonyl-glycine *p*-nitrophenyl ester [13] (3.55 gm, 12 mmol) and imidazole [11] (1.2 g about 18 mmol) are added and the suspension is stirred at room temperature overnight. The resin is thoroughly

[9] washed on a filter with toluene and then with dichloromethane and dried in air. The i.r. spectrum reveals a strong ester carbonyl band at 1730 cm^{-1}. Amino acid analysis of a hydrolysate shows a "loading" of about 1.2 mmol per g resin [14].

Benzyloxycarbonyl-L-isoleucyl-Polymer

A sample of the hydroxymethyl-resin (1.0 g, about 2 meq.) is suspended in toluene (7 ml) and treated with imidazole (0.41 g, about 6 mmol) and with benzyloxycarbonyl-L-isoleucine *p*-nitrophenyl ester [15] (1.54 g, 4 mmol). The mixture is stirred at room temperature overnight and worked up as described in the preceding paragraph. Amino acid analysis of a hydrolysate indicates about 0.8 mmol isoleucine per g of resin. No alloisoleucine is present [16, 17].

1. Merrifield RB (1963) J Amer Chem Soc 85: 2149
2. Bodanszky M, Fagan DT (1977) Int J Peptide Protein Res 10: 375
3. Bodanszky M, Sheehan JT, Chem & Ind 1966: 1597
4. Beyerman HC, in't Veld RA (1969) Recueil 88: 1019
5. Schreiber J, in Peptides 1966 (Beyerman HC et al., eds.). North Holland Publ. Amsterdam: 1967, p. 107
6. Commercially available.
7. Meaning the product of copolymerization of a mixture of 99% styrene and 1% divinylbenzene or 98% styrene and 2% divinylbenzene.
8. A small sample is placed on the tip of an improvised spatula made from a thick copper wire and held in the upper part of the colorless flame of a Bunsen burner. Halogens produce a bright green color. The color is more readily observed in a semi-dark room.
9. All washings should be carried out slowly to allow time for the diffusion of solutes from the inside of the resin particles.
10. The time needed for completion of saponification is a function of the physical properties of the resin and can range from a few hours to several days. If complete saponification is not achieved in a day the reaction can be facilitated by the addition of dichloromethane (20 ml), which enhances the swelling of the polymer.
11. Stewart FHC (1968) Austr J Chem 21: 1639
12. The resin swells in toluene (Khan SA, Sivanandaiah KM, Chem Commun 1976: 614). In more polar solvents, racemization might take place.
13. Sandrin E, Boissonnas RA (1963) Helv Chim Acta 46: 1637
14. If it is desirable to block the remaining alcoholic hydroxyl groups this can be done by acetylation with acetic anhydride in toluene.
15. Bodanszky M, du Vigneaud V (1959) J Amer Chem Soc 81: 5688
16. Bodanszky M, Conklin LE, Chem Commun 1967: 773
17. Imidazole catalyzed transesterification is usually accompanied by racemization, but not in toluene.

2.4 4-Nitrobenzyl Esters [1, 2]

L-Alanine 4-Nitrobenzyl Ester *p*-Toluenesulfonate [1]

$$H_2N{-}CH(CH_3){-}COOH + HOCH_2{-}C_6H_4{-}NO_2 + CH_3{-}C_6H_4{-}SO_3H \xrightarrow{-H_2O}$$

$$CH_3{-}C_6H_4{-}SO_3H \cdot H_2N{-}CH(CH_3){-}CO{-}OCH_2{-}C_6H_4{-}NO_2$$

$C_{17}H_{20}N_2O_7S$ (396.4)

L-Alanine (8.9 g, 100 mmol), 4-nitrobenzyl alcohol (76.5 g, 500 mmol) and *p*-toluenesulfonic acid (monohydrate, 57 g, 300 mmol) are suspended in dry chloroform [3] (300 ml). The suspension is heated to boiling: soon a clear solution forms. The chloroform is distilled, the distillate is dried with anhydrous $CaSO_4$ (Drierite) and returned into the distillation flask [4]. This continuous removal of water formed in the esterification reaction [5] leads to the complete disappearance of L-alanine from the mixture in about six hours [6]. The solvent is removed in vacuo and the residue triturated with ether. The crude ester, *p*-toluenesulfonate, is collected on a filter, washed with ether [7] and dried in air. The yellow-orange material is crystallized from isopropanol (250 ml). The purified product (36.4 g, 92%) is pale yellow and melts at 155 to 157.5 °C [8]. It is chromatographically homogeneous.

1. Mazur RH, Schlatter JM (1963) J Org Chem 28: 1025. The paper describes the preparation of 4-nitrobenzyl esters of several more amino acids.
2. 4-Nitrobenzyl esters were prepared also through benzyloxycarbonylamino acids (Schwarz H, Arakawa K (1959) J Amer Chem Soc 81: 5691
3. The chloroform should be freed from water and alcohol by shaking with concentrated sulfuric acid and then filtered through a filter aid such as Celite or Hyflo-Supercel.
4. A continuously working apparatus can be improvised for this purpose or the dried distillate must be returned from time to time into the flask, preferably through an attached dropping funnel.
5. The water stems, in part, from the dehydration of *p*-toluenesulfonic acid.
6. On thin layer chromatograms no alanine can be detected with ninhydrin.
7. A major portion (about 42 g) of 4-nitrobenzyl alcohol can be recovered from the ether filtrate and washings. These are combined, extracted with 1 N $KHCO_3$ and evaporated to dryness in vacuo. The residue is triturated with benzene, filtered, washed with benzene and dried.
8. The m.p. can be raised to 156–158 °C by a second recrystallization from isopropanol: $[\alpha]_D^{29}$ $-4°$ (*c* 1, methanol).

2.5 4-Methoxybenzyl Esters of N^{α}-Protected Amino Acids [1]

Benzyloxycarbonylglycine 4-Methoxybenzyl Ester [2]

$C_6H_5-CH_2O-CO-NH-CH_2-C(=O)-O-C_6H_4-NO_2 + HOCH_2-C_6H_4-OCH_3 \xrightarrow{\text{imidazole}}$

$C_6H_5-CH_2O-CO-NH-CH_2-CO-OCH_2-C_6H_4-OCH_3 + HO-C_6H_4-NO_2$

$C_{18}H_{19}NO_5$ (329.3)

A solution of benzyloxycarbonylglycine 4-nitrophenyl ester [3] (3.30 g, 10 mmol) and 4-methoxybenzyl alcohol [4] (1.38 g = 1.25 ml, 10 mmol) in dioxan [5] (40 ml) is treated with imidazole (6.8 g, 100 mmol). After about 3 hours at room temperature the mixture is diluted with water (about 200 ml) and extracted with ether [6] (300 ml in three portions). The extract is washed with N NH_4OH, water, 0.1 N HCl and again with water (100 ml each), dried over $MgSO_4$ and evaporated to dryness in vacuo. The residue, an oil, solidifies on cooling. The ester, 2.5 g (76%) melts after recrystallization from ethyl acetate-cyclohexane at 55.5–56 °C [7].

1. Weygand F, Hunger K (1962) Chem Ber 95: 1; cf. also McKay FC, Albertson NF (1957) J Amer Chem Soc 79: 4686
2. Stewart FHC (1968) Austral J Chem 21: 2543
3. Bodanszky M, du Vigneaud V (1962) Biochem Prep 9: 110
4. Anise alcohol, *p*-anisyl alcohol.
5. In the case of chiral amino acids recemization can occur in more polar solvents such as dimethylformamide. The use of strongly basic catalysts also affects the chiral purity of the products.
6. In the preparation of peptide esters ethyl acetate was used (cf. ref. 2).
7. When the same compound was obtained by condensing benzyloxycarbonylglycine and 4-methoxybenzyl alcohol with the aid of dicyclohexylcarbodiimide (ref. 1) a melting point of 60 °C was observed. Esterification with the help of dimethylformamide dineopentyl acetal (Brechbühler H, Büchi H, Hatz E, Schreiber J, Eschenmoser A (1965) Helv Chim Acta 48: 1746 gave a product melting at 61.5–62.2 °C.

2.6 Benzhydryl Esters (Diphenylmethyl Esters) [1, 2]

$$CH_3-C_6H_4-SO_3H \cdot H_2N-CHR-COOH + \bar{N}=\overset{+}{N}=C(C_6H_5)_2 \longrightarrow$$

$$CH_3-C_6H_4-SO_3H \cdot H_2N-CHR-CO-OCH(C_6H_5)_2 + N_2$$

p-Toluenesulfonic acid (monohydrate, 20 g, 105 mmol) is added to a solution (or suspension) of the amino acid (100 mmol) in water (50 to 100 ml). The solution [3] is evaporated to dryness in vacuo and the residue is washed with acetone. The *p*-toluenesulfonate salts are recrystallized from hot acetone [4].

The amino acid *p*-toluenesulfonate (10 mmol) is dissolved in dimethylformamide (5 ml) and the stirred solution is warmed to 50 °C. A solution of diazodiphenylmethane [5] (15 mmol) in dimethylformamide (10 ml) is added and stirring is continued at 50 °C for ten more minutes. The mixture is cooled to room temperature and the solvent is removed in vacuo. The esters are recrystallized from acetonitrile. Yields range from 75 to 90% [6]. The products give satisfactory values in elemental analysis.

1. Aboderin AA, Delpierre GR, Fruton JS (1965) J Amer Chem Soc 87: 5469
2. Benzhydryl esters have also been prepared through the reaction of diphenylchloromethane with silver or triethylammonium salts of *N*-acylamino acids (Stelakatos GC, Paganou A, Zervas L, J Chem Soc (C) 1966: 1191). Direct esterification with diphenylmethanol has also been described (Hiskey RG, Adams JB, Jr. (1965) J Amer Chem Soc 87: 3969). Benzhydryl

esters are cleaved by alkaline hydrolysis, by HCl in nitromethane or in ethyl acetate, HBr in nitromethane and also by hydrogenolysis.

3. Prolonged stirring or mild heating may be necessary to obtain a clear solution.
4. The salts of glycine and of L-alanine were recrystallized [1] from water-acetone-ether.
5. Smith LI, Howard KL Organic Syntheses, Coll Vol. III, Wiley and Sons, New York, 1955, p. 351
6. In the cases of L-arginine and L-histidine the di-β-naphthalenesulfonate salts were used. The yields on esters were 78 and 53% respectively.

2.7 Phthalimidomethyl Esters

N-Chloromethylphthalimide [2] (as precursor)

$$\text{C}_6\text{H}_4(\text{CO})_2\text{NH} + \text{CH}_2\text{O} \longrightarrow \text{C}_6\text{H}_4(\text{CO})_2\text{N-CH}_2\text{-OH} \xrightarrow{\text{SOCl}_2} \text{C}_6\text{H}_4(\text{CO})_2\text{N-CH}_2\text{-Cl}$$

$C_9H_6O_2NCl$ (195.7)

Phthalimide (147.1 g, 1 mol) is added to a 20% solution of formaldehyde in water containing 1.1 mol (33 g) formaldehyde. The suspension is slowly heated to reflux temperature until all the phthalimide dissolves. On cooling crystals of N-hydroxymethylphthalimide separate. The crystals are collected on a filter, thoroughly washed with water and with 95% ethanol and dried in vacuo over sulfuric acid. The product, about 165 g (93%) melts at 141 °C. The hydroxymethyl derivative is suspended in thionyl chloride (about 200 ml) and the suspension slowly heated to reflux temperature and refluxed until all the solid material dissolves. After refluxing for an additional 30 min, the excess thionyl chloride is removed in vacuo and the residue recrystallized from toluene. The chloromethyl derivative (140 g, 72% based on phthalimide) melts at 132 °C. It should be stored protected from moisture, in a well closed bottle.

Esterification of Protected Amino Acids

$$\text{C}_6\text{H}_5\text{-CH}_2\text{O-CO-NH-CHR-COO}^- \ \text{H}_2\overset{+}{\text{N}}(\text{C}_2\text{H}_5)_2 + \text{Cl-CH}_2\text{-N}(\text{CO})_2\text{C}_6\text{H}_4 \longrightarrow$$

$$\text{C}_6\text{H}_5\text{-CH}_2\text{O-CO-NH-CHR-CO-OCH}_2\text{-N}(\text{CO})_2\text{C}_6\text{H}_4 + (\text{C}_2\text{H}_5)_2\text{NH}\cdot\text{HCl}$$

The protected amino acid (or peptide) is dissolved in ethyl acetate containing an equivalent amount of diethylamine. One mole of *N*-chloromethylphthalimide is added and the mixture stored at 37–40 °C overnight. The solution is washed twice with water and twice with a 2% solution of $NaHCO_3$ in water. The organic layer is dried over $MgSO_4$ and the solvent removed in vacuo to leave the usually crystalline product [3–5].

1. Nefkens GHL, Tesser GI, Nivard RJF (1963) Rec Trav Chim Pay-Bas 82: 941
2. Pucher GW, Johnson TB (1922) J Amer Chem Soc 44: 817
3. Melting points and specific rotations of the *N*-hydroxymethylphthalimide esters of several benzyloxycarbonylamino acids are reported in ref. 1.
4. The phthalimidomethyl esters of benzyloxycarbonylamino acids were hydrogenated in methanol containing an equimolecular amount of *p*-toluenesulfonic acid. After the removal of the catalyst and the solvent the *p*-toluenesulfonate salts of amino acid *N*-hydroxymethylphthalimide esters were crystallized from ethanol-ether.
5. Esters of *N*-hydroxymethylphthalimide are cleaved by a saturated solution of HCl in dioxane at room temperature, overnight and also by HBr in acetic acid, in 15 minutes. Removal of the carboxyl protecting group can be achieved also with sodium hydroxide in aqueous ethanol.

2.8 *tert*-Butyl Esters

Addition of N^{α}-Acylamino Acids to Isobutylene [1]

L-Proline *tert*-Butyl Ester

$$C_6H_5-CH_2O-CO-N-CH-COOH + CH_3-C(CH_3)=CH_2 \xrightarrow{H_2SO_4}$$

$$C_6H_5-CH_2O-CO-N-CH-CO-O-C(CH_3)_2-CH_3 \xrightarrow{H_2/Pd}$$

$$HN-CH-CO-O-C(CH_3)_2-CH_3 + CO_2 + C_6H_5-CH_3$$

$C_9H_{17}NO_2$ (171.2)

To a solution of benzyloxycarbonyl-L-proline (24.9 g, 100 mmol) in dichloromethane (200 ml) concentrated sulfuric acid (1 ml) is added and the mixture saturated with isobutylene. This results in an increase of the volume to about 300 ml. The flask is stoppered and kept at room temperature for 3 days. After the addition of a solution of Na_2CO_3 (10 g) in water (150 ml) the two layers are separated and the organic phase washed with distilled water (three times, 10 ml each time). The solvent is removed in vacuo at a bath temperature of 60 °C. The residue, first an oil, solidifies in crystalline form. It weighs 28.8 g (94%) and melts at 44–45 °C.

The entire amount of benzyloxycarbonyl-L-proline *tert*-butyl ester thus prepared is dissolved in absolute ethanol (250 ml) and hydrogenated in the presence of a 10% Pd-on-charcoal catalyst (3 g). When no more CO_2 evolves [2] the catalyst is removed by filtration and the solution evaporated [3] in vacuo. The residue is dissolved in ether (150 ml), extracted [4] with a 10%

solution of Na_2CO_3 in water (twice, 50 ml each time), dried over anhydrous $MgSO_4$ and evaporated [3] to dryness. The *tert*-butyl ester (12.5 g, 73% [5]) is on oil which boils at 57 °C (1.5 mm). It can be stored for prolonged periods of time.

Transesterification with *tert*-Butyl Acetate [6]

$$H_2N{-}CHR{-}COOH + CH_3CO{-}O{-}C(CH_3)_3 \xrightarrow{HClO_4} H_2N{-}CHR{-}CO{-}O{-}C(CH_3)_3 + CH_3COOH$$

The amino acid (10 mmol), *tert*-butyl acetate (150 ml) and perchloric acid (11 meq $HClO_4$ in the form of a 60% aqueous solution) are mixed and shaken at room temperature until a clear solution is formed. This requires about 15 minutes. The solution is stored at room temperature for four days [7]. The mixture is cooled in an ice-water bath and extracted with 0.5 N HCl (four times, 25 ml each time). The aqueous extracts are immediately neutralized with solid $NaHCO_3$. The combined aqueous solutions are extracted with ether (4 times, 100 ml each time), the ether extracts pooled, dried over $MgSO_4$ and the ether removed by distillation [3] in vacuo. Addition of a solution of dry HCl in ether precipitates the amino acid *tert*-butyl ester hydrochloride in crystalline form. Yields range from 40 to 70% [8].

Esterification of Free Amino Acids through Acid Catalyzed Addition to Isobutylene [9]

L-Tyrosine *tert*-Butyl Ester

$$HO{-}C_6H_4{-}CH_2{-}CH(NH_2){-}COOH + CH_3{-}C(CH_3){=}CH_2 \xrightarrow{H_2SO_4} HO{-}C_6H_4{-}CH_2{-}CH(NH_2){-}CO{-}O{-}C(CH_3)_3$$

$C_{13}H_{19}NO_3$ (237.3)

Liquid isobutylene (30 ml) is slowly added to L-tyrosine (3.62 g, 20 mmol) dissolved in a mixture of dioxane (30 ml) and *p*-toluenesulfonic acid (monohydrate [10], 7.6 g, 40 mmol) in a 500 ml pressure bottle [11]. The flask is securely stoppered, wrapped in a towel and shaken at room temperature for about 20 hours. The solution is poured into an ice-cold mixture of ethyl acetate (120 ml) and 0.25 N NaOH (120 ml). The pH is adjusted to 9.1 [12] and the *tert*-butyl ester extracted into ethyl acetate (twice, 200 ml each time). The solvent is removed in vacuo [3]. The crystalline residue [13] (2.15 g, 45%) melts at 140–143 °C; $[\alpha]_D^{25}$ +24.4 (*c* 2, EtOH). The m.p. can be raised to 143–145 °C by recrystallization from ethyl acetate-hexane.

1. Anderson GW, Callahan FM (1960) J Amer Chem Soc 82: 3359
2. If the escaping gas gives no precipitate when bubbled through a half saturated solution of $Ba(OH)_2$ in water, the reaction is complete.

3. Since *tert*-butyl esters of most amino acids are sufficiently volatile to be partially lost, during the removal of solvents, distillation at only moderately reduced pressure and the use of a distillation column can be recommended.
4. Extraction with a solution of sodium carbonate in water removes some unchanged acylamino acid.
5. Calculated on benzyloxycarbonyl-L-proline.
6. Taschner E, Chimiak A, Bator B, Sokolowska T (1961) Justus Liebigs Ann Chem 646: 134
7. With some amino acids the perchlorate of the *tert*-butyl ester slowly separates from the solution.
8. A very modest yield was obtained in the esterification of proline.
9. Roeske, R. (1963) J Org Chem 28: 1251
10. Instead of *p*-toluenesulfonic acid concentrated sulfuric acid (3 ml) can be used.
11. Larger pressure bottles are less reliable.
12. For esters of other amino acids a higher pH might be preferable. For instance, ph 9.5 was recommended [9] in the case of *N*-tosyl-L-lysine. Tyrosine ester, however, is soluble in alkali.
13. Most amino acids yield oily *tert*-butyl esters. These can be converted to crystalline salts, e.g. to the hydrochloride by the addition of HCl in ether to their ethereal solutions.

2.9 Phenacyl Esters [1, 2]

Benzyloxycarbonylglycine Phenacyl Ester [3]

$$C_6H_5-CH_2O-CO-NH-CH_2-COOH + N(C_2H_5)_3 + BrCH_2-CO-C_6H_5 \longrightarrow$$

$$C_6H_5-CH_2O-CO-NH-CH_2-CO-OCH_2-CO-C_6H_5 + HBr\cdot N(C_2H_5)_3$$

$C_{18}H_{17}NO_5$ (327.3)

Triethylamine (1.01 g = 1.4 ml, 10 mmol) and phenacyl bromide [4] (2.0 g, 10 mmol) are added to a solution of benzyloxycarbonylglycine (2.1 g, 10 mmol) in ethyl acetate (20 ml). The reaction mixture is allowed to stand at room temperature overnight. The crystalline precipitate, triethylammonium bromide, which separated during the reaction, dissolves on the addition of water (20 ml). More ethyl acetate (20 ml) is added and the two layers are separated. The organic phase is washed with water (20 ml), 0.5 M $KHCO_3$ (20 ml) and again with water (20 ml) and dried over anhydrous $MgSO_4$. The solvent is removed in vacuo and the residue triturated with petroleum ether (b.p. 40–60 °C). The crystals are collected on a filter, washed with petroleum ether and dried in air. Recrystallization from isopropanol yields analytically pure benzyloxycarbonylglycine phenacyl ester (2.7 g, 83%) melting at 103 °C.

1. Stelakatos GC, Paganou A, Zervas L, J Chem Soc (C) 1966: 1191
2. Phenacyl esters are cleaved by sodium thiophenoxide (Sheehan JC, Daves GD, Jr. (1964) J Org Chem 29: 2006) and also by hydrogenolysis (cf. ref. 1).
3. The same procedure is applicable for the preparation of the phenacyl esters of Z-L-Val, Z-L-Ala, Z-L-Phe, Z-L-Leu, and the diester of Z-L-Asp as well.
4. α-Bromoacetophenone. It is a potent lachrymator.

2.10 2-Trimethylsilylethyl Esters [1]

$$C_6H_5-CH_2O-CO-NH-CHR-COOH + HO-CH_2-CH_2-Si(CH_3)_3 + C_6H_{11}N{=}C{=}NC_6H_{11} \longrightarrow$$

$$C_6H_5-CH_2O-CO-NH-CHR-CO-OCH_2-CH_2Si(CH_3)_3 + C_6H_{11}NH-CO-NHC_6H_{11}$$

A stirred solution of the N^{α}-protected amino acid (10 mmol) in acetonitrile [2] (6–10 ml) is treated with pyridine (1.6 ml) and 2-trimethylsilylethanol [3] (1.72 ml = 1.42 g, 12 mmol) and cooled in an ice water bath for ten minutes. Dicyclohexylcarbodiimide (2.3 g, 11 mmol) is added and the mixture stirred in the ice water bath for about one hour and then kept in the refrigerator overnight. A 5 M solution of oxalic acid in dimethylformamide (0.3 ml) is added. About 1/2 hour later the precipitate (dicyclohexylurea) is removed by filtration and washed on the filter with ethyl acetate. The combined filtrate and washings [4] are extracted with dilute hydrochloric acid, with a solution of sodium hydrogen carbonate, dried over $MgSO_4$ and evaporated to dryness in vacuo. The residue is chromatographed on a column of silica gel with ethyl acetate-hexane as eluent. The fractions which contain the desired ester in homogeneous form are pooled and the protected amino acid 2-trimethylsilylethyl ester is secured [5], by the removal of the solvents in vacuo.

1. Sieber P (1977) Helv Chim Acta 60: 2711
2. If the protected amino acid is not sufficiently soluble in acetonitrile, dissolution can be achieved by the addition of a slight amount of dimethylformamide.
3. Speier JL, Webster JA, Barnes GH (1957) J Amer Chem Soc 79: 974; Gerlach H (1977) Helv Chim Acta 60: 3039. The alcohol is commercially available.
4. Ethyl acetate must be used in sufficient amount to produce two layers.
5. The esters of (Z)-Gln, (Z)-Phe, (Z)-Trp, (Z)-Thr(But), (Z)-Tyr(But) were obtained in crystalline form. Most of the esters, however, remained as oils. Physical constants of a series of protected amino acid 2-trimethylsilylethyl esters are described in ref. 1.

2.11 4-Picolyl Esters

Benzyloxycarbonyl-L-phenylalanine 4-Picolyl Ester [1]

$$C_6H_5-CH_2O-CO-NH-CH(CH_2C_6H_5)-COOH + ClCH_2-C_5H_4N\cdot HCl + 2\,(CH_3)_2N-C(=NH)-N(CH_3)_2 \longrightarrow$$

$$C_6H_5-CH_2O-CO-NH-CH(CH_2C_6H_5)-CO-O-CH_2-C_5H_4N + 2\,(CH_3)_2N-C(=NH_2\cdot Cl)-N(CH_3)_2$$

$C_{23}H_{22}N_2O_4$ (390.4)

A solution of benzyloxycarbonyl-L-phenylalanine (3.0 g, 10 mmol), 4-picolyl chloride hydrochloride [2] (1.64 g, 10 mmol) and tetramethylguanidine [3] (2.3 g = 2.5 ml, 20 mmol) in dimethylformamide (25 ml) is heated at 90 °C for 2 hours [4]. The solvent is removed in vacuo (ca 100 Pa), ethyl acetate (about 25 ml) is added, removed in vacuo and the addition and removal of ethyl acetate is repeated. The residue is dissolved in ethyl acetate (30 ml), the solution is washed three times with N $NaHCO_3$ (10 ml each time), twice with water (10 ml each time), once with a saturated solution of sodium chloride (10 ml) and dried over $MgSO_4$. After evaporation of the solvent in vacuo the residue is crystallized from ether with the application of activated charcoal and recrystallized similarly. The colorless ester (1.9 g, 49%) melts at 87.5–89.5 °C; $[\alpha]_D^{20} - 33°$ (*c* 1, dimethylformamide). On elemental analysis the calculated values are found. The nmr spectrum ($CDCl_3$) shows the expected resonances.

Benzyloxycarbonyl-L-valine 4-Picolyl Ester [1]

$$C_6H_5-CH_2O-CO-NH-CH(CH(CH_3)_2)-COOH + HO-CH_2-C_5H_4N + C_6H_{11}-N{=}C{=}N-C_6H_{11} \longrightarrow$$

$$C_6H_5-CH_2O-CO-NH-CH(CH(CH_3)_2)-CO-O-CH_2-C_5H_4N + C_6H_{11}-NH-CO-NH-C_6H_{11}$$

$C_{19}H_{22}N_2O_4$ (342.4)

To a solution of 4-(hydroxymethyl)pyridine [5] (1.09 g, 10 mmol) and benzyloxycarbonyl-L-valine (2.76 g, 11 mmol) in dichloromethane (25 ml) dicyclohexylcarbodiimide (2.27 g, 11 mmol) is added and the mixture is stirred at room temperature overnight. The separated *N,N'*-dicyclohexylurea is removed by filtration and the combined filtrate and washings extracted, twice, with N $NaHCO_3$ (30 ml each time) and once with water (10 ml). The solvent is removed in vacuo, the residue is extracted into 2 N solution of citric acid in water (75 ml in three portions) and the solution is made alkaline by the addition of solid $NaHCO_3$. The product is extracted into ethyl acetate, the solution dried over $MgSO_4$ and evaporated to dryness. Recrystallization of the residue from ether-petroleum ether (b.p. 40–60 °C) yields the 4-picolyl ester (2.0 g; 59%) m.p. 63–65 °C; $[\alpha]_D^{20} - 11.6°$ (*c* 1, dimethylformamide).

1. Camble R, Garner R, Young GT, J Chem Soc (C) 1969: 1911
2. 4-Chloromethyl-pyridine hydrochloride. Commercially available.
3. The use of triethylamine instead of tetramethylguanidine led to lower yields.
4. Completion of the reaction is indicated by the disappearance of 4-picolyl chloride from the mixture. This can be revealed on thin layer plates of silicagel in the solvent system chloroform-methanol (19:1).
5. (4-Pyridyl)-methanol. Commercially available.

2.12 9-Fluorenylmethyl Esters [1, 2]

***tert*-Butyloxycarbonyl-L-phenylalanine 9-Fluorenylmethyl Ester [1]**

$C_{28}H_{29}NO_4$ (443.5)

9-(Hydroxymethyl-)fluorene [3] (2.15 g, 11 mmol) and imidazole [4] (1.02 g, 15 mmol) are added to toluene [5] (20 ml). The suspension is stirred at room temperature and *tert*-butyloxycarbonyl-L-phenylalanine *o*-nitrophenyl ester [6, 7] (3.86 g, 10 mmol) is added. Stirring is continued overnight. The solvent is removed in vacuo, the residue dissolved in ethyl acetate (150 ml) and the solution washed twice with a 2% solution of citric acid in water (100 ml each time), water (100 ml), 0.5 N $NaHCO_3$ solution (100 ml) and again with water (100 ml). The ethyl acetate solution is dried over $MgSO_4$ and evaporated to dryness. The residue is crystallized from ethyl acetate-hexane to yield 3.1 g (70%) ester melting at 125–126 °C; $[\alpha]_D^{23} -6°$ (*c* 1, $CHCl_3$) [7]. On tlc the material moves as a single spot, readily detected by its strong uv absorption: R_f 0.73 (in ethyl acetate-hexane 3:7). In the NMR spectrum ($CDCl_3$) the expected resonances are observed. On elemental analysis correct values are obtained for C, H and N.

1. Bednarek MA, Bodanszky M (1983) Int J Peptide Protein Res 21: 196
2. Kessler H, Siegmeier R, Tetrahedron Letters, 1983: 281
3. 9-Fluorenylmethanol. Commercially available.
4. Stewart FHC (1968) Austr J Chem 21: 1639
5. Toluene is dried by distillation: water is removed with the foreruns. In more polar solvents racemization might occur.
6. Bodanszky M, Funk KW, Fink ML (1973) J Org Chem 38: 3565
7. Other active esters, such as *p*-nitrophenyl and 2,4,5-trichlorophenyl esters are also suitable for the preparation of 9-fluorenylmethyl esters by transesterification.

2.13 2-Benzyloxyphenyl Esters [1, 2]

L-Alanine 2-Benzyloxyphenyl Ester Hydrochloride [3]

$(CH_3)_3C-O-CO-NH-CH(CH_3)-COOH + Cl-C(=O)-OC_2H_5 + N(C_2H_5)_3 \longrightarrow$

$(CH_3)_3C-O-CO-NH-CH(CH_3)-CO-O-C(=O)-OC_2H_5 + (C_2H_5)_3N \cdot HCl \xrightarrow{\text{2-(OH)}C_6H_4\text{-O-CH}_2\text{-}C_6H_5}$

$(CH_3)_3C-O-CO-NH-CH(CH_3)-CO-O-C_6H_4-O-CH_2-C_6H_5 + CO_2 + C_2H_5OH \xrightarrow{HCl}$

$HCl \cdot H_2N-CH(CH_3)-CO-O-C_6H_4-O-CH_2-C_6H_5$

$C_{16}H_{18}NO_3Cl$ (307.8)

To a solution of *tert*-butyloxycarbonyl-L-alanine (1.89 g, 10 mmol) in chloroform (15 ml) triethylamine (1.01 g = 1.4 ml, 10 mmol) is added. The solution is cooled to about −5 °C and stirred during the addition of ethyl chlorocarbonate (1.09 g = 0.96 ml, 10 mmol) and also during the subsequent addition of a precooled (about −5 °C) solution of 2-benzyloxyphenol [4] (2.00 g, 10 mmol) and triethylamine (1.01 g = 1.4 ml, 10 mmol) in chloroform (10 ml). The addition of this solution should take 5 to 10 minutes. The mixture is kept at about −5 °C for 30 minutes and then at room temperature for about 6 hours. The solvent is removed in vacuo and the residue dissolved by the addition of water (50 ml) and ethyl acetate (50 ml). The two phases are separated and the ethyl acetate solution washed with a 10% solution of citric acid in water (40 ml). The solution is dried over anhydrous $MgSO_4$ and evaporated in vacuo. The residue, an oil (ca. 2.7 g) is dissolved in ether (25 ml), and treated with an about 3 N solution of HCl in ethyl acetate (2.5 ml). After standing at room temperature overnight the crystalline hydrochloride is collected on a filter, washed with ether and dried in vacuo over NaOH pellets. The product (1.95 g, 64%) melts at 164–171 °C; $[\alpha]_D^{20}$ −15.7° (*c* 1, $CHCl_3$).

1. Jones JH, Young GT, J Chem Soc, (C) 1968: 436
2. Removal of the benzyl group (e.g., by catalytic hydrogenation) converts this protecting group into an activating group: 2-hydroxyphenyl esters are acylating agents.
3. Cowell RD, Jones JH, J Chem Soc Perkin I 1972: 2236
4. Catechol monobenzyl ether. Cf. ref. 1 and also J Druey (1935) Bull Soc Chim France 52: 1737. Commercially available.

2.14 Phenyl Esters [1]

$$(CH_3)_3C-O-C(=O)-NH-CH(CH_2C_6H_5)-COOH \; + \; HO-C_6H_5 \xrightarrow{\text{BOP–Reagent}} (CH_3)_3C-O-C(=O)-NH-CH(CH_2C_6H_5)-C(=O)-O-C_6H_5$$

***tert*-Butyloxycarbonyl-L-phenylalanine Phenyl Ester [2]**

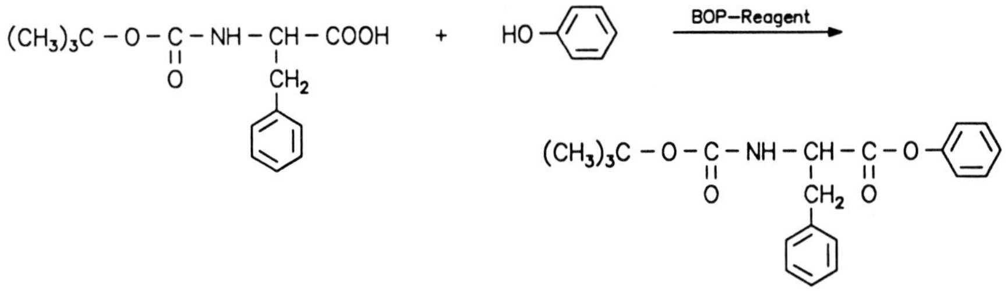

$C_{20}H_{23}NO_4$ (341.4)

To a solution of *tert*-butyloxycarbonyl-L-phenylalanine (2.65 g, 10 mmol) in dichloromethane (30 ml), phenol (0.94 g, 10 mmol) is added followed by the addition of triethylamine (2.02 g, 2.8 ml, 10 mmol) and of benzotriazol-1-yloxytris-(dimethylamino)phosphonium) hexafluorophosphate (4.42 g, 10 mmol) (3). The reaction mixture is kept at room temperature for two hours. After the addition of a saturated aqueous solution of sodium chloride (150 ml), the mixture extracted with ethyl acetate, the extract washed with 2 N HCl (three times), with a saturated solution of sodium bicarbonate (three times) and with a saturated sodium chloride solution (twice), dried over anhydrous magnesium sulfate and evaporated in vacuo to dryness. The oily residue (3.2 g, 94%) spontaneously crystallizes and then melts at 91 °C.

1. Phenyl esters can be cleaved by peroxide catalyzed hydrolysis with alkali, under very mild conditions. (Kenner GW, Seely JH (1972) J Amer Chem Soc 94: 3259
2. Castro B, Evin G, Selve C, Seyer R (1977) Synthesis 413
3. "BOP-reagent" (Castro B, Dormoy JR, Dourtoglou B, Evin G, Selve C, Synthesis 1976: 751. Commercially available. It is mainly used for the formation of the peptide bond (cf. p. 124).

2.15 Allyl Esters [1, 2]

$$C_6H_5-CH_2O-C(=O)-NH-CH(CH_2OH)-COOH \; + \; Br-CH_2-CH=CH_2 \xrightarrow{NaHCO_3} C_6H_5-CH_2O-C(=O)-NH-CH(CH_2OH)-C(=O)-O-CH_2-CH=CH_2$$

Benzyloxycarbonyl-L-serine Allyl Ester [1]

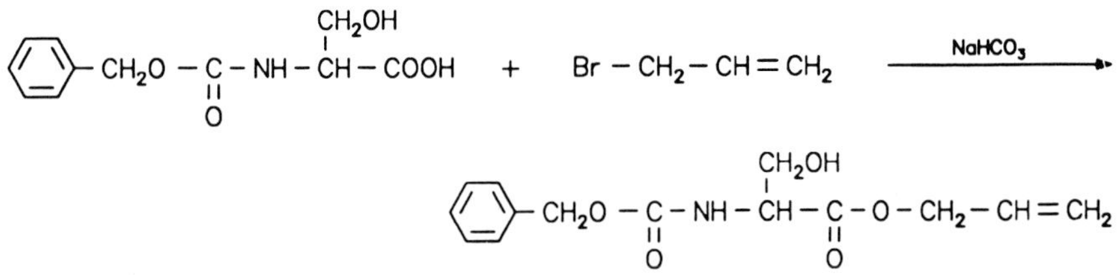

$C_{14}H_{17}NO_5$ (279.3)

Benzyloxycarbonyl-L-serine (2.4 g, 10 mmol) is dissolved in a 5% solution of sodium bicarbonate in water (15 ml). A solution of tricaprylmethylammonium chloride ("aliquat 336", 4.0 g, 10 mmol) and allyl bromide (6.0 g, 10 mmol) in dichloromethane (15 ml) is added. The mixture is vigorously stirred at room temperature for 72 hours, then extracted with dichloromethane (three times, 25 ml each time). The extracts are combined, dried over anhydrous magnesium sulfate and the solvent removed in vacuo. The oily residue is purified by chromatography on a silica gel column with a 2:1 mixture of ether and ethyl acetate as eluent. The chromatographically homogeneous ester, a colorless oil, weighs 2.31 g (83%), $[\alpha]_D^{22} + 5°$ (*c* 1, $CHCl_3$) CO bands at $1770\ cm^{-1}$ and $1730\ cm^{-1}$. The expected resonances are present in the nmr spectrum and elemental analysis gives the calculated values.

1. Friedrich-Bochtnitschek S, Waldmann H, Kunz H (1989) J Org Chem 54: 751
2. In the presence of a Pd(O) catalyst, the allyl ester group is cleaved by transfer to morpholine as accepting nucleophile.

3 Protection of Side Chain Functions

3.1 Serine Ethers

N-tert-Butyloxycarbonyl-O-benzyl-L-serine Cyclohexylammonium Salt [1, 2]

$$CH_3-C(CH_3)_2-O-CO-NH-CH(CH_2OH)-COOH \xrightarrow{2\,NaH} CH_3-C(CH_3)_2-O-CO-NH-CH(CH_2ONa)-COONa \xrightarrow[2.\ HCl\quad 3.\ C_6H_{11}NH_2]{1.\ BrCH_2-C_6H_5}$$

$$CH_3-C(CH_3)_2-O-CO-NH-CH(CH_2O-CH_2-C_6H_5)-COOH\cdot H_2N-C_6H_{11}$$

$C_{15}H_{21}NO_5\cdot C_6H_{13}N$ (394.5)

A sample of *N-tert*-butyloxycarbonyl-L-serine (20.5 g, 100 mmol) is dissolved in dimethylformamide [3] (500 ml), the solution is cooled to 0 °C and treated with sodium hydride [4] (8.2 g 65% material, 220 mmol). When no more gas evolves benzyl bromide [5] (18.8 g = 13.1 ml, 110 mmol) is added and the mixture is stirred at 25–30 °C for about 5 hours. The solvent is removed in vacuo at 40 °C bath temperature, the residue dissolved in water (500 ml) and the solution extracted with ether (twice, 200 ml each time). The aqueous layer is acidified to pH 3.5 with 3 N HCl and then extracted with ethyl acetate (five times, each time with 200 ml). The extracts are pooled, washed with water (twice, 200 ml each time) and dried over anhydrous $MgSO_4$. The solvent is removed in vacuo and the residue, a colorless oil, dissolved in ether (300 ml). Cyclohexylamine (9.0 g = 10.4 ml, ca 90 mmol) is added to the solution: the cyclohexylammonium salt of *N-tert*-butyloxycarbonyl-*O*-benzyl-L-serine precipitates. It is collected on a filter and thoroughly washed with ether. After recrystallization from ethyl acetate the product (about 22 g, 56%) melts at 159–160 °C; $[\alpha]_D^{25}$ −29° (*c* 1, methanol). On elemental analysis values close to the ones calculated for C, H and N, are obtained.

1. Sugano H, Miyoshi M (1976) J Org Chem 41: 2352
2. The procedure described here is a modification of an earlier proposed method, which gave somewhat lower yields (Hruby VJ, Ehler KW (1970) J Org Chem 35: 1690
3. Dimethylformamide is dried over a molecular sieve and then over NaH.
4. The contact of water with sodium hydride should be carefully avoided.
5. Because of the evolution of hydrogen and the lachrimatory effect of benzyl bromide these operations must be carried out in a well ventilated hood.

3.2 Ethers of Threonine

***O-tert*-Butyl-L-threonine [1, 2]**

$$C_6H_5\text{-}CH_2O\text{-}COCl + H_2N\text{-}CH(CHOH\text{-}CH_3)\text{-}COOH \xrightarrow[2.\ HCl]{1.\ Na_2CO_3} C_6H_5\text{-}CH_2\text{-}O\text{-}CO\text{-}NH\text{-}CH(CHOH\text{-}CH_3)\text{-}COOH$$

$$\xrightarrow[N(C_2H_5)_3]{BrCH_2\text{-}C_6H_4\text{-}NO_2} C_6H_5\text{-}CH_2O\text{-}CO\text{-}NH\text{-}CH(CHOH\text{-}CH_3)\text{-}CO\text{-}OCH_2\text{-}C_6H_4\text{-}NO_2 \xrightarrow[H_2SO_4]{CH_3\text{-}C(CH_3)=CH_2}$$

$$C_6H_5\text{-}CH_2O\text{-}CO\text{-}NH\text{-}CH(CHO\text{-}C(CH_3)_3\text{-}CH_3)\text{-}CO\text{-}OCH_2\text{-}C_6H_4\text{-}NO_2 \xrightarrow{H_2/Pd} H_2N\text{-}CH(CHO\text{-}C(CH_3)_3\text{-}CH_3)\text{-}COOH$$

$C_8H_{17}NO_3$ (175.2)

A solution of L-threonine (119.1 g, 1 mol) in 2 N NaOH (500 ml) is stirred and cooled in an ice-water bath. Benzyl chlorocarbonate [3] (171 g = 144 ml, 1 mol) and 2 N Na_2CO_3 (750 ml) are added alternately, each in about ten approximately equal portions. Stirring is continued for about 30 min., then the alkaline solution is extracted with ether (twice, 250 ml each time). Concentrated hydrochloric acid is added to the cooled and stirred aqueous layer until it is acid to Congo. The oil which separates is transferred into ethyl acetate (500 ml), the aqueous solution saturated with NaCl and extracted with ethyl acetate (500 ml). The organic extracts are pooled, washed with water (500 ml), with a saturated solution of NaCl in water (500 ml) and dried over anhydrous Na_2SO_4. The solvent is removed in vacuo and the residue crystallized from ethyl acetate-hexane. The product (about 240 g, 95%) melts at 101–102 °C; $[\alpha]_D^{20} -6°$ (*c* 2, AcOH).

An aliquot (50.6 g, 200 mmol) of benzyloxycarbonyl-L-threonine dissolved in ethyl acetate (200 ml) is treated with triethylamine (30.3 g = 42 ml, 300 mmol) and with *p*-nitrobenzyl bromide (64.8 g, 300 mmol). The mixture is kept in a bath of 80 °C for about 8 to 9 hours and then cooled to room temperature. The separated triethylammonium bromide is removed by filtration and the solution washed with 2 N HCl (200 ml), water (200 ml), 10% $NaHCO_3$ (200 ml), again with water (200 ml), dried over anhydrous Na_2SO_4 and evaporated in vacuo. The residue is dissolved in a small volume of ethyl acetate and diluted with hexane till turbid. When crystallization is complete the product is collected, washed with a mixture of ethyl acetate and hexane and dried: 75.4 g (97%); m.p. 114–115 °C $[\alpha]_D^{20}$ −14.0° (*c* 2, methanol) [4].

A 100 mmol aliquot of the *p*-nitrobenzyl ester (38.8 g) is dissolved in dichloromethane (400 ml), the solution is placed in a thick-walled round bottom flask and cooled in an ice-water bath. Isobutylene (350 ml) and concentrated sulfuric acid (5 ml) are added with caution. The flask is stoppered, wrapped in a towel and kept at room temperature for four days. The mixture is cooled again to about 0 °C, washed with an ice-cold 5% solution of Na_2CO_3 in water (3 times, 200 ml each time). The aqueous extracts are reextracted with dichloromethane (twice, 100 ml time), the organic solutions pooled and washed with water until the washes are neutral. The solution is dried over P_2O_5 and evaporated in vacuo. The yellow, crystalline residue is dissolved in a small volume of ethyl acetate, diluted with hexane until some dark oil separates.The clear solution is decanted and further diluted with hexane. The *tert*-butyl ether separates in needles (36 g, 81%) melting at 55–56.5 °C. The product is analytically and chromatographically pure; R_f 0.50 (in heptane-*tert*-butanol-pyridine, 5:1:1).

The fully blocked amino acid (22.3 g, 50 mmol) is dissolved in methanol (150 ml), the solution diluted with water (50 ml) and acetic acid (4 ml), the air displaced with nitrogen, a 10% Pd on charcoal catalyst (4.5 g) is added and the mixture hydrogenated at room temperature and atmospheric pressure. After removal of the catalyst by filtration (under nitrogen) and the solvent by evaporation in vacuo the residue is triturated with ethanol and recrystallized from methanol-acetone. The product (7.5 g, 85%) is analytically pure [6]. It melts with decomposition at 259–260 °C; $[\alpha]_D^{20}$ −42.1° (*c* 2, methanol).

1. Wünsch E, Jentsch J (1964) Chem Ber 97: 2490
2. The same procedure is applicable for the preparation of *O-tert*-butyl-L-serine and *O-tert*-butyl-L-tyrosine as well.
3. Benzyl chloroformate; carbobenzoxychloride. If the reagent is not freshly prepared a larger amount may be necessary. The acid chloride content of the liquid can be determined in a small scale experiment by the acylation of excess glycine.
4. Catalytic hydrogenation yields enantiomerically pure L-threonine.
5. The catalyst might be pyrophoric and should be disposed of with care.
6. Treatment of a sample with trifluoroacetic acid for 6 hours at room temperature followed by evaporation of the trifluoroacetic acid and precipitation of L-threonine with triethylamine in acetone afforded enantiomerically pure material.

3.3 Tyrosine Ethers

***O*-Benzyl-L-tyrosine [1, 2]**

$C_{16}H_{17}NO_3$ (271.3)

L-Tyrosine (18.1 g, 100 mmol) is dissolved in 2 N NaOH (100 ml) [3] and a solution of cupric sulfate (pentahydrate, 12.5 g, 50 mmol = 100 meq) [4] in water (50 ml) is added. A precipitate forms and soon dissolves. The mixture is heated to 60 °C, cooled to room temperature, diluted with methanol (350 ml) and made more alkaline with 2 N NaOH (15 ml). This is followed by the addition of benzyl bromide [5] (17.1 g = 12 ml, 100 mmol). The mixture is vigorously stirred at 25–30 °C for about one and a half hours. The purple-blue precipitate is collected on a filter, washed with a mixture of methanol (50 ml) and water (175 ml), then with methanol (25 ml) and dried in air [6].

The well disintegrated copper complex is triturated and washed [7] with N HCl (5 times, 50 ml each time), with distilled water (twice, 25 ml each time), with approximately 1.5 N NH_4OH (5 times, 25 ml each time) and finally with water (twice, 25 ml each time). The crude product weighs about 17.5 g (64%), melts with decomposition between 260 and 270 °C; $[\alpha]_D^{20}$ −9.5 (*c* 1, 80% AcOH) [8].

1. Wünsch E, Fries G, Zwick A (1958) Chem Ber 91: 542.
2. The procedure described here is a modification of the published [1] method.
3. In ref. 1 one equivalent of NaOH is recommended for the dissolution of one mole tyrosine. To obtain a clear solution, however, almost two equivalents are needed.

4. An equivalent amount of cupric sulfate seems to be sufficient. A large excess was applied in ref. 1.
5. Benzyl bromide (or α-bromotoluene) is one of the worst lachrimators. The operations described above should be carried out in a well ventilated hood.
6. Alternatively, one can proceed without drying the precipitate.
7. Trituration and washing with HCl, water and NH_4OH are conveniently carried out on a sinter-glass filter with the help of a sturdy glass rod provided with a flattened head. The solutions containing cupric chloride and ammonium chloride are removed by suction.
8. Recrystallization from a large volume of boiling 80% acetic acid does not raise the m.p. and leaves the specific rotation unchanged.

3.4 Acylation of the ε-Amino Group of Lysine

$N^ε$-Trifluoroacetyl-lysine [1, 2]

Ethyl thioltrifluoroacetate [3]

$$(CF_3CO)_2O + CH_3CH_2SH \longrightarrow CF_3\overset{O}{\overset{\|}{C}}-SC_2H_5 + CF_3COOH$$

$C_4H_5F_3OS$ (158.1)

These operations must be carried out in a well ventilated hood.

Ethyl mercaptane (62.1 g=74 ml, 1 mol) is cooled in an ice-water bath and trifluoroacetic anhydride (262 g=177 ml, 1.25 mol) is added dropwise with intermittent shaking. About one hour is required for the addition of the anhydride. The mixture is allowed to stand under a reflux condenser, provided with a $CaCl_2$-filled drying tube, for about an hour and then heated at 100 °C for 3 hours. The reddish solution is cooled to room temperature, washed with a 5% solution of KOH in water (twice, 1 liter each time), with water (twice, with one liter each time) dried over anhydrous $MgSO_4$ and distilled at atmospheric pressure. The fraction boiling at 88–90 °C is collected. It weighs about 95 g (60%), d_4^{20} 1.25; n_D^{20} 1.374.

Trifluoroacetylation of lysine [1, 2, 4]

$$CF_3\overset{O}{\overset{\|}{C}}-SC_2H_5 + H_2N-\underset{}{\overset{NH_2-(CH_2)_4}{CH}}-COOH \longrightarrow H_2N-\overset{CF_3CO-NH-(CH_2)_4}{CH}-COOH + CH_3CH_2SH$$

$C_8H_{13}F_3N_2O_3$ (242.2)

Lysine monohydrochloride [2] (18.3 g, 100 mmol) is dissolved in N NaOH (100 ml) and treated with ethyl thioltrifluoroacetate (25 g=20.0 ml, 158 mmol). The heterogeneous mixture is shaken at room temperature for 6 hours. A precipitate appears and gradually turns the solution into a mass of crystals. The mixture is cooled in an ice-water bath, filtered and the product washed with ice-water (about 50 ml in several portions). The crude material (about 18 g) is dissolved in boiling water (100 ml) and the solution diluted with hot ethanol (150 ml). On cooling crystals separate. These are collected on a filter, washed with 60% ethanol (50 ml) and dried in air. The purified product (12.5 g,

52%) melts with decomposition at 226–231 °C [5, 6]. A second recrystallization affords an analytical sample.

1. Schallenberg, EE, Calvin M (1955) J Amer Chem Soc 77: 2779; cf. also Greenstein GP, Winitz M (1961) Chemistry of the Amino Acids, Wiley and Sons, New York, p. 915.
2. In reference 1 the preparation of N^{ε}-trifluoroacetyl-DL-lysine is described. Since the chiral center is not affected by the reaction it is reasonable to assume that the procedure is applicable for the synthesis of N^{ε}-trifluoroacetyl-L-lysine as well.
3. Hauptschein M, Stokes CS, Nodiff EA (1952) J Amer Chem Soc 74: 4005. Commercially available as *S*-ethyl trifluorothioacetate.
4. Trifluoroacetylation with the thiol ester should be carried out in a well ventilated hood: the ethyl mercaptane, set free in the reaction, must be allowed to escape from the reaction vessel.
5. Yield and melting point are those of the derivative of DL-lysine.
6. The preparation of an N^{ε}-substituted lysine derivative without the aid of a copper complex or some other means of blocking the α-amino function would seem to indicate considerable regio-selectivity in the acylation reaction with ethyl thioltrifluoroacetate. Yet, the same procedure is applicable [1] for the trifluoroacetylation of the amino group of glycine and norleucine as well. Thus, it is more likely that also some N^{α}-substituted product forms in the reaction but remains in solution. The N^{ε}-blocked derivatives of lysine are usually fairly insoluble in water while the N^{α}-acyl derivatives are more readily soluble. For instance N^{ε}-benzyloxycarbonyl-L-lysine was obtained (in moderate yield) by the acylation of the amino acid with benzyl chlorocarbonate. The alkaline reaction mixture was neutralized with HCl and the precipitate thoroughly washed with water. Most of the N^{α}-benzyloxycarbonyl derivative remains in solution and the rest is removed during washing with water (M. Bodanszky, unpublished).

N^{ε}-*p*-Toluenesulfonyl-L-lysine [1]

Lysine $\xrightarrow{CuCO_3Cu(OH)_2}$ [$NH_2(CH_2)_4CH(NH_2)COO]_2Cu$ $\xrightarrow[CH_3C_6H_4SO_2Cl]{NaHCO_3}$ [$CH_3C_6H_4SO_2NH(CH_2)_4CH(NH_2)COO]_2Cu$

$\xrightarrow{H_2S}$ CuS + 2 $CH_3C_6H_4SO_2NH(CH_2)_4CH(NH_2)COOH$

$C_{13}H_{20}N_2O_4S$ (300.4)

A solution of L-lysine monohydrochloride (18.3 g, 100 mmol) in water (1 liter) is heated to reflux while cupric carbonate (basic, 30 g) is added with caution. After two hours of boiling the undissolved cupric carbonate is removed from the hot mixture by filtration and washed with hot water (100 ml). The combined filtrate and washings are cooled to room temperature and treated with sodium hydrogen carbonate (32 g). The mixture is vigorously stirred and a solution of *p*-toluenesulfonyl chloride (28.7 g, 150 mmol) in acetone (1 liter) is added. Stirring is continued overnight. The precipitated light blue copper complex is collected on a filter, thoroughly washed with water, acetone and ether and dried in air. It weighs about 23–25 g and melts at 238–240 °C with decomposition [2]. The finely powdered copper complex (23 g, 70 mmol) is suspended in boiling distilled water (400 ml) and a stream of H_2S is passed through the solution [3] for 30 min. Boiling is continued to remove excess H_2S [4]. Hydrochloric acid (6 N, 12 ml) is added and the mixture filtered from CuS [5]. The pH of the filtrate is adjusted to about 6 with 4 N NaOH: the product, N^{ε}-tosyl-L-lysine is fairly insoluble in water and separates from the solution. The mixture cooled to room temperature, the product is collected on a filter, washed first with water, then with ethanol and dried in air: 16.4 g (55%); m.p. 233–234 °C dec. A sample purified by dissolution in dilute HCl and precipitation with dilute NaOH melts at 237–238 °C dec.; $[\alpha]_D$ +13.6 (*c* 3, 2 N HCl). On elemental analysis correct values are obtained for C, H, N and S.

1. Roeske R, Stewart FHC, Stedman RJ, du Vigneaud V (1956) J Amer Chem Soc 78: 5883; cf. also Erlanger BF, Sachs H, Brand E (1954) J Amer Chem Soc 76: 1806 for the preparation of the next homolog of N^{ε}-tosyl-L-lysine, namely N^{δ}-tosyl-L-ornithine.
2. The temperature should be raised by 3 °C per minute during the determination of the melting point.
3. A well ventilated hood is necessary for this operation: H_2S is highly toxic.
4. Decomposition of copper complexes is possible also without the use of H_2S. To a suspension of the copper complex (10 mmol) in water (50 ml) thioacetamide (1.12 g, 15 mmol) is added, then 2 N NaOH to bring the pH of the suspension to 8. The mixture is stirred at room temperature for one day. The pH is lowered to 1.6 by the addition of 2 N HCl and the cupric sulfide removed by filtration. The filtrate is neutralized to precipitate the product. (Taylor UF, Dyckes DF, Cox JR, Jr (1982) Internat J Peptide Protein Res. 19: 158).
5. To overcome difficulties usually experienced in the filtration of CuS, activated charcoal (about 3 g) is added with caution (to avoid overboiling). The addition of filter-aid, such as celite (also about 3 g) facilitates the filtration and further improvement can be achieved by pre-coating the filter with a 1:1 mixture of charcoal and celite.

3.5 Protection of the Guanidino Group of Arginine

Nitro-L-arginine [1]

$CH_2-NH-C(=NH)-\overset{+}{N}H_3$ / CH_2 / CH_2 / $H_2N-CH-COO^-$ $\xrightarrow[H_2SO_4]{HNO_3}$ $CH_2-NH-C(=N\cdot NO_2)-NH_2$ / CH_2 / CH_2 / $H_3\overset{+}{N}-CH-COO^-$

$C_6H_{13}N_5O_4$ (219.2)

A mixture of fuming sulfuric acid containing 30% SO_3 (15 ml) and fuming nitric acid (23 ml) is prepared in a round bottom flask surrounded by an ice-salt bath. L-Arginine free base (17.4 g, 100 mmol) is added to the well stirred mixture through a powder funnel in small portions. The material on the funnel is rinsed into the flask with concentrated sulfuric acid (8 ml) and stirring is continued at low temperature for an additional hour. The mixture is then poured in a thin stream [2] onto cracked ice (about 200 g) and concentrated ammonium hydroxide is added to the solution, with stirring, to a pH of 8 to 9. The pH is readjusted to 6 with a small amount of acetic acid and the resulting suspension is stored in the cold overnight. The product is collected on a filter, washed with cold water and recrystallized from boiling water. The purified material is washed on a filter with 95% ethanol and dried in air. It weighs about 18 g (82%) and melts at 251–252 °C with decomposition [3]; $[\alpha]_D^{25}$ +24° (*c* 4, 2 N HCl) [4]. On elemental analysis correct values are obtained for C, H and N [5].

1. Hofmann K, Peckham WD, Rheiner A (1956) J Amer Chem Soc 78: 238.
2. Handling of the corrosive mixture requires proper caution.
3. Nitro-L-arginine isolated from nitrated proteins (Kossel A, Kenneway EL (1911) Hoppe Seyler's Z Physiol Chem 72: 486 melted at 227–228 °C, while the same material prepared by nitration of arginine nitrate salt was found (Bergmann M, Zervas L, Rinke H (1934) Hoppe Zeyler's Z Physiol Chem 224: 40 to have a (corrected) m.p. of 263 °C (dec.).
4. Nitroarginine and nitroarginine containing peptides can be readily detected on thin layer plates by their absorption in the u.v. The same absorption (ε ca 14000 at 265 nm) is useful also in the determination of their concentration.
5. Slow combustion might be necessary to obtain correct values for nitroarginine and its derivatives.

N^{α}-Benzyloxycarbonyl-N^{G}-*p*-toluenesulfonyl-L-arginine [1]

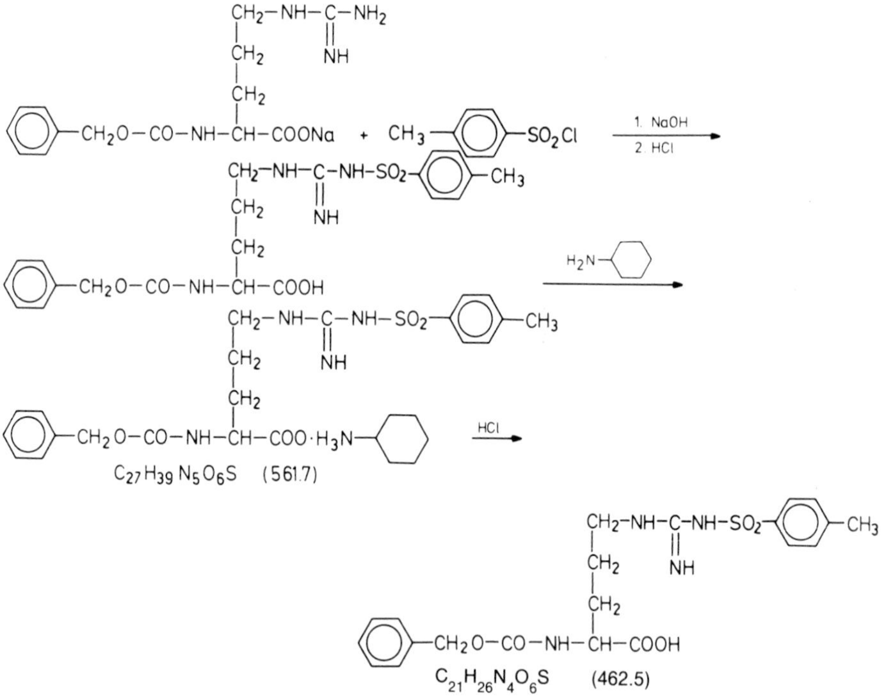

1. NaOH
2. HCl

H_2N–C₆H₁₁ (cyclohexylamine)

HCl

$C_{27}H_{39}N_5O_6S$ (561.7)

$C_{21}H_{26}N_4O_6S$ (462.5)

A suspension of *N*-benzyloxycarbonyl-L-arginine [2] (30.8 g, 100 mmol) in a mixture of distilled water (125 ml) and acetone (500 ml) is cooled in an ice-water bath. Precooled 4 N NaOH is added with vigorous stirring to bring the pH to 11–11.5 and to maintain it in this range. The starting material dissolves in about two hours. A solution of *p*-toluenesulfonyl chloride (47.5 g, 250 mmol) in acetone (75 ml) is added dropwise to the well-stirred solution for a period of about 30 min while the pH is kept at 11–11.5 by the addition of (precooled) 4 N NaOH to the cooled reaction mixture. Stirring is continued in the cold for about 3 more hours. The solution is neutralized (to pH 7) with N HCl and the acetone is removed in vacuo at a bath temperature of 20 to 25 °C. The solution is diluted with water (250 ml) and extracted three times with ether (200 ml each time). The aqueous layer is acidified to pH 3 by the addition of 6 N HCl with stirring and cooling: a viscous oil separates. The solution is decanted from the oil, saturated with sodium chloride and extracted three times with ethyl acetate (200 ml each time). The oil and the ethyl acetate extracts are combined, washed with 0.1 N HCl until the washes give no Sakaguchi test [3] and then with water until the washes are neutral. The organic layer is dried over anhydrous sodium sulfate, the solvent removed in vacuo. The viscous oily product (37 g, ca. 80 mmol) is dissolved in methanol (125 ml), the solution cooled in an ice-water bath and treated with cyclohexylamine (8.4 g = 9.7 ml, ca. 82 mmol, a small excess over the amount of the protected amino acid determined by the weight of the oily product). The mixture is diluted with ether to cloudiness and a few drops of methanol are added to produce a clear solution. On scratching the wall of the vessel with a glass rod crystals form which are collected after storage in the refrigerator for two days. They are washed with ice-cold methanol and dried: the cyclohexylammonium salt weighs about 27 g (48%), melts at 152–154 °C; $[\alpha]_D^{25}$ 6.1° (*c* 3.2, methanol). Recrystallization from methanol-ether yields an analytical sample with unchanged melting point.

A sample of the cyclohexylammonium salt (28.1 g, 50 mmol) is dissolved in methanol (200 ml) with warming, the solution cooled with ice-water and treated with 2 N HCl (40 ml, 80 mmol). Stirring is continued for an hour. The methanol is removed in vacuo at a bath temperature of 20 °C, the remaining solution diluted with water (250 ml) and extracted four times with ethyl acetate (each time 200 ml). The ethyl acetate extracts are pooled, washed with water until the wash is neutral, dried over anhydrous Na_2SO_4 and evaporated in vacuo to dryness. The residue is redissolved in ethyl acetate (200 ml) with warming and the solution is cooled. The crystals are collected on a filter, washed with a small amount of ethyl acetate and dried in air: 21 g (91% calculated on the amount of cyclohexylammonium salt), m.p. 86–89 °C; $[\alpha]_D^{25}$ −1.3° (*c* 4, dimethylformamide). On elemental analysis correct values are obtained for C, H and N.

1. Ramachandran J, Li CH (1962) J Org Chem 27: 4006.
2. Boissonnas RA, Guttmann S, Huguenin RL, Jaquenoud PA, Sandrin E (1958) Helv Chim

Acta 41: 1867; Zervas L, Winitz M, Greenstein JP (1962) J Org Chem 26: 3348; cf. also Bergmann M, Zervas L (1932) Ber dtsch Chem Ges 65: 1192. The preparation of this compound is described in the present volume on p. 12.

3. A spot of the wash on filter paper is dried, sprayed with a solution of 0.1% α-naphthol in 0.1 N NaOH. After drying a second spray is applied, prepared from ice-cold 5% KOH (100 ml) and bromine (0.07 ml). Arginine appears as a red spot.

N[α]-Benzyloxycarbonyl-*N*[δ],*N*[ω]-bis-(1-adamantyloxycarbonyl)-L-arginine [1]

$C_{36}H_{48}N_4O_8$ (664.8)

A solution of *N*[α]-benzyloxycarbonyl-L-arginine [2] (30.8 g, 100 mmol) in dioxane (60 ml) and 2 N NaOH (200 ml) is cooled to 6–8 °C and stirred vigorously [3]. Next 1-adamantyl chlorocarbonate [4] (86 g, 400 mmol) in dioxane (75 ml) and 2 N NaOH (300 ml) are added simultaneously over about one hour, dropwise. Stirring [3] at 6–8 °C is continued for three or more hours. The solid material is collected by centrifugation, triturated with ether, filtered and washed with ether. Most of the solvent is removed in vacuo and the residue triturated with hexane. The precipitate is filtered, washed with hexane and dried in air. This fraction and the ether insoluble material are combined, suspended in water and acidified to pH 2–3 with 0.5 M citric acid. The acid thus liberated is transferred into ether and the ether solution dried over anhydrous Na_2SO_4. The solvent is removed in vacuo and the foamy residue crystallized from methanol-water. The product (about 60 g, ca 90%) melts at 120–122 °C dec.; $[\alpha]_D^{22}$ +20.8° (*c* 1, $CHCl_3$). On elemental analysis correct values are found for C, H and N.

1. Jäger G, Geiger R (1970) Chem Ber 103: 1727.
2. Boissonnas RA, Guttmann S, Huguenin RL, Jaquenoud PA, Sandrin E (1958) Helv Chim Acta 41: 1867. The preparation of *N*[α]-benzyloxycarbonyl-L-arginine is described in this volume on p. 12.
3. A vibrator was used for this purpose by the authors of ref. 1.
4. Haas WL, Krumkalns EV, Gerzon K (1966) J Amer Chem Soc 88: 1988. The preparation of the chlorocarbonate is described on page 197 of this volume.

N^{α}-4-Methoxy-benzyloxycarbonyl-N^{G}-mesitylene-sulfonyl-L-arginine Cyclo-hexylammonium Salt [1, 2]

$$H_3CO-C_6H_4-CH_2O-CO-NH-CH(COOH)-(CH_2)_3-NH-C(=NH)-NH_2 + H_3C-C_6H_2(CH_3)_2-SO_2Cl \xrightarrow{1.\ NaOH;\ 2.\ HCl}$$

$$\xrightarrow{H_2N-C_6H_{11}} H_3CO-C_6H_4-CH_2O-CO-NH-CH(COO^-)-(CH_2)_3-NH-C(=N-SO_2-C_6H_2(CH_3)_3)-NH_2 \quad H_3\overset{+}{N}-C_6H_{11}$$

$C_{30}H_{45}N_5O_7S$ (619.8)

A solution of mesitylene-2-sulfonyl chloride (45 g, ca 200 mmol) in acetone (200 ml) is added, dropwise, over a period of about 40 minutes, to a stirred mixture of 4-methoxybenzyloxycarbonyl-L-arginine (34 g, 100 mmol), 4 N NaOH (100 ml) and acetone (500 ml), cooled in an ice-bath. Stirring is continued for two additional hours, then the solvent removed in vacuo at a bath temperature of 30 °C. The residue is dissolved in water, the solution washed with ethyl acetate, acidified with citric acid and the separated material extracted into ethyl acetate. The extract is washed with ice cold 0.2 N HCl, with a solution of sodium chloride, dried over anhydrous sodium sulfate and evaporated in vacuo. The oily residue is dissolved on acetone and treated with cyclohexylamine (10 g, 11.5 ml, 100 mmol). The crystalline material is collected on a filter and recrystallized (twice) from methanol-acetonitrile. The purified salt (38 g, 61%) melts at 125–128 °C, $[\alpha]_D^{22} + 5.5°$ (*c* 0.7, MeOH). It gives a single spot on thin layer chromatograms and the expected values on elemental analysis.

1. Yajima H, Takeyama M, Kanaki J, Nishimura O (1978) Chem Pharm Bull 26: 3752
2. Transfer of the mesitylenesulfonyl group to the side chain of tyrosine residues is less pronounced than in the use of the *p*-toluenesulfonyl group. The application of scavangers (anisole, thioanisole, *o*-cresol) is still indicated.

N^{G}-4-Methoxy-2,3,6-trimethyl-benzenesulfonyl-L-arginine [1, 2]

4-Methoxy-2,3,6-trimethyl-benzenesul-fonylchloride (precursor)

$$H_3CO-C_6H_2(CH_3)_3 + HO-SO_2-Cl \longrightarrow H_3CO-C_6H(CH_3)_3-SO_2Cl \quad \text{(I)}$$

$C_{10}H_{13}O_3SCl$ (248.7)

A solution of 2,3,5-trimethylanisole (15.1 g, 100 mmol) in dichloromethane (500 ml) is cooled to −5 to −10 °C while a solution of chlorosulfonic acid (20 ml, 35 g, ca 300 mmol) in dichloromethane (300 ml) is added with stirring. Stirring is continued for three hours. The mixture is then poured onto crushed ice (250 ml) containing 5% sodium bicarbonate. The organic layer is separated, washed with water, dried over anhydrous magnesium sulfate and the solvent removed in vacuo. The residue is crystallized from *n*-hexane. The pure sulfonic acid chloride weighs 18 g (72%); m.p. 56–58 °C.

N^{α}-benzyloxycarbonyl-N^{G}-4-methoxy-2,3,6-trimethylbenzenesulfonyl-L-arginine Cyclohexylammonium Salt

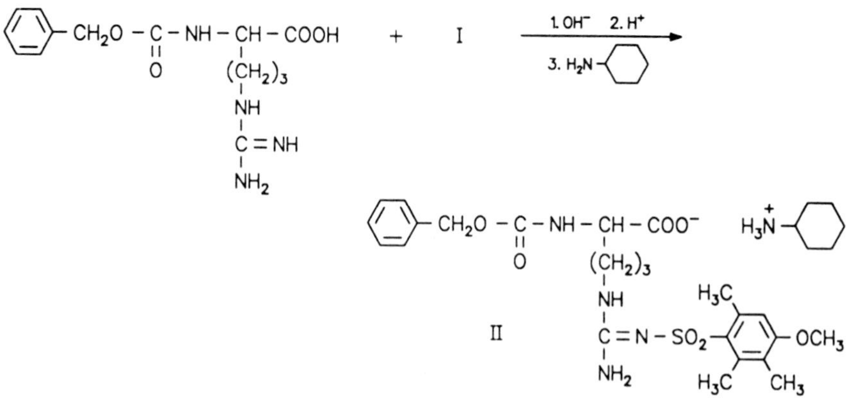

$$C_6H_5-CH_2O-C(=O)-NH-CH(COOH)-(CH_2)_3-NH-C(=NH)-NH_2 \; + \; \mathrm{I} \xrightarrow[3.\ H_2N-C_6H_{11}]{1.\ OH^-\quad 2.\ H^+}$$

$$C_6H_5-CH_2O-C(=O)-NH-CH(COO^-)-(CH_2)_3-NH-C(=N-SO_2-C_6H(CH_3)_3-OCH_3)-NH_2 \quad H_3\overset{+}{N}-C_6H_{11}$$

II

$C_{30}H_{45}N_5O_7S$ (619.8)

A solution of *N*-benzyloxycarbonyl-L-arginine (3.1 g, 10 mmol) in a mixture of 4 N NaOH (10 ml) and acetone (40 ml) is cooled in an ice-water bath and treated with a solution of the above described sulfonyl chloride (compound I, 4.35 g, 17.5 mmol) in acetone (10 ml). The mixture is stirred at room temperature for three hours and then acidified with a 10% solution of citric acid in water. The solvent is evaporated, the residue dissolved in ethyl acetate (30 ml) and cyclohexylamine (0.99 g, 1.15 ml, 10 mmol) is added. The crystalline precipitate is collected on a filter and recrystallized from methanol-ethyl acetate. The analytically pure salt weighs 4.45 g (72%) and melts with decomposition at 195–197 °C; $[\alpha]_D^{23}$ +6.5° (*c* 1.2, MeOH).

N^{G}-4-methoxy-2,3,6-trimethylbenzenesulfonyl-L-arginine

$$\mathrm{II} \xrightarrow[2.\ H_2/Pd]{1.\ H^+} H_2N-CH(COOH)-(CH_2)_3-NH-C(=N-SO_2-C_6H(CH_3)_3-OCH_3)-NH_2$$

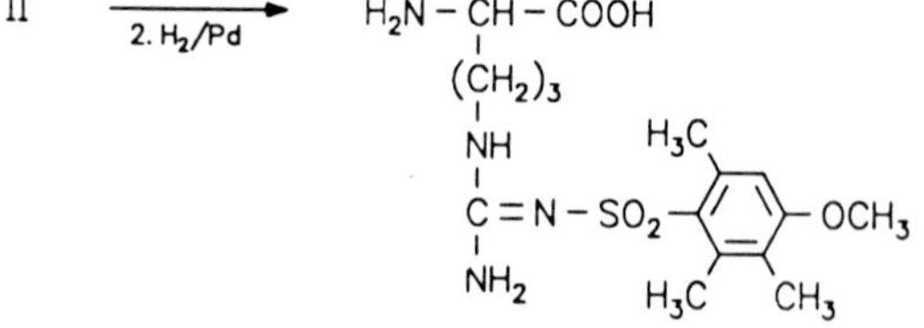

$C_{16}H_{26}N_4O_5S \cdot 1/2\ H_2O$ (395.5)

A suspension of the cyclohexylamine salt (compound II, 3.1 g, 5 mmol) in ethyl acetate (70 ml) is shaken with 0.2 N H_2SO_4 (30 ml), the phases are separated and the organic layer evaporated to dryness. The residue is dissolved in methanol and hydrogenated in the presence of a Pd catalyst. The product obtained after the removal of the catalyst and the solvent [3] is crystallized from water: 1.56 g (79%). It melts at 100–103 °C, $[\alpha]_D^{23}$ −4.9° (*c* 1.3, MeOH). The analytical values indicate a hemihydrate.

1. Fujino M, Wakimasu M, Kitada C (1981) Chem Pharm Bull 29: 2825.
2. The 4-methoxy-2.3.6-trimethylbenzenesulfonyl group is cleaved by a solution of thioanisole in trifluoroacetic acid.
3. Removal of the catalyst requires special care: in contact with Pd catalysts, methanol is readily ignited by air. Therefore, filtration from the catalyst should be carried out under a blanket of nitrogen or carbon dioxide. For removal of the benzyloxycarbonyl group by hydrogenolysis cf. pages 129–134.)

3.6 Masking the Imidazole in Histidine [1]

3.6.1 N^{im}-*p*-Toluenesulfonyl-L-histidine

N^{α}-Benzyloxycarbonyl-N^{im}-tosyl-L-histidine [2]

1. $CH_3-C_6H_4-SO_2Cl$
2. dil H_2SO_4
3. $(C_6H_{11})_2NH$

$C_{33}H_{44}N_4O_6S$ (624.8)

A solution of N^{α}-benzyloxycarbonyl-L-histidine [3] (14.5 g, 50 mmol) and sodium carbonate (10.6 g, 100 mmol) in water (150 ml) is kept at 10–15 °C while *p*-toluenesulfonyl chloride (tosyl chloride, 12.8 g, 67 mmol) is added in small portions with vigorous stirring. Stirring is continued at room temperature for four hours. The unreacted acid chloride is extracted with ether (twice, 75 ml each time) and the aqueous solution acidified to pH 2 with 1 N H_2SO_4. An oil separates. It is transferred into ethyl acetate (150 ml) and the solution is reextracted twice more with ethyl acetate (150 ml each time). The organic extracts are pooled, washed with water (300 ml), dried over anhydrous Na_2SO_4 and the solvent removed in vacuo. The residue is redissolved in ethyl acetate (150 ml) and treated with dicyclohexylamine (8.8 ml, 8.0 g, 44 mmol). The crystalline dicyclohexylammonium salt is collected on a filter, washed with ethyl acetate (50 ml) and dried in air. Recrystallization from methanol-ethyl acetate yields 22 g (70%) of the product melting at 150–152 °C dec., $[\alpha]_C$ +19° (*c* 1, dimethylformamide). It is analytically pure.

N^{im}-*p*-Toluenesulfonyl-L-histidine [2]

Z-L-His(Tos).DCHA —(1 H_2SO_4; 2 HBr/AcOH)→ $H_2N-CH(CH_2\text{-Im-}SO_2\text{-}C_6H_4\text{-}CH_3)-COOH$

$C_{13}H_{15}N_3O_4S$ (309.3)

The dicyclohexylammonium salt described in the preceding paragraph (6.25 g, 10 mmol) is suspended in 1 N H_2SO_4 (40 ml) and the suspension shaken with ethyl acetate (50 ml). The aqueous phase is reextracted with ethyl acetate (50 ml) and the pooled extracts dried over anhydrous Na_2SO_4. The solvent is removed in vacuo and the residue is treated with ca. 4 N HBr in acetic acid. After 1 hour at room temperature the mixture is diluted with dry ether (120 ml). The ether solution is decanted from the insoluble material, the latter is thoroughly washed, by trituration and decantation with ether and dried in vacuo over NaOH pellets. It is then dissolved in methanol (60 ml) and the solution neutralized with pyridine (about 4 ml). The precipitated N^{im}-tosyl-L-histidine [4, 5] is collected on a filter and thoroughly washed with methanol. The dry product (2.2 g, 71%) melts at 140–145 °C.

1. Sakakibara S, Fujii T (1969) Bull Chem Soc Japan 42: 1466
2. Fujii T, Sakakibara S (1974) Bull Chem Soc Japan 47: 3146
3. Patchornik A, Berger A, Katchalski E (1957) J Amer Chem Soc 79: 6416; Akabori S, Okawa K, Sakiyama F (1958) Nature 181: 772; Sakiyama F, Okawa K, Yamakawa T, Akabori S (1958) Bull Chem Soc Japan 31: 926
4. Throughout this section the τ rather than π position appears to be substituted by the tosyl group. Yet, this is based only on steric preference and proposed without proof.
5. This intermediate is suitable for the preparation of *N*-blocked derivatives. It can be dissolved in aqueous $NaHCO_3$ and treated e.g., with *tert*-butyl azidocarbonate. The N^{im}-tosyl group is resistant to moderately strong acids and requires HF for its removal. On the other hand the tosyl group is cleaved from the imidazole nucleus by nucleophiles. Thus it can migrate to free α-amino groups and is displaced by 1-hydroxybenzotriazole with the formation of 1-tosyloxybenzotriazole [2]. The related N^{im}-*p*-methoxysulfonyl group (Kitagawa K, Kitade K, Kiso Y, Akita T, Funakoshi S, Fujii N, Yajima H. J Chem Soc Chem Commun 1979: 955) is cleaved by trifluoroacetic acid in the presence of dimethyl sulfide at room temperature in about an hour.

3.6.2 N^α-*tert*-Butyloxycarbonyl-N^π-benzyloxymethyl-L-histidine [1]

N^α-,N^π-Di-*tert*-Butyloxycarbonyl-L-histidine Methyl Ester

$Cl^- \; H_3\overset{+}{N}-CH(CH_2\text{-Im(N-H)})-COOCH_3$ —($[(CH_3)_3C-O-CO]_2O$)→ $(CH_3)_3C-O-C(=O)-NH-CH(CH_2\text{-Im(N-}C(=O)-O-C(CH_3)_3))-COOCH_3$

I

First, triethylamine (28 ml, 200 mmol), then di-*tert*-butyl pyrocarbonate [2] (48 g, 220 mmol) are added to a suspension of L-histidine methyl ester dihydrochloride (24.2 g, 100 mmol) in methanol (80 ml) and the mixture is stored at room temperature overnight. The solvent is removed in vacuo and chloroform (250 ml is added to the residue. The solution is extracted with a 10% solution of citric acid in water (twice, with 20 ml each time), dried over anhydrous sodium sulfate and evaporated in vacuo. The oily residue is converted to a solid by trituration with light petroleum ether (b.p. 40–60 °C). The product (33 g, 90%) melts at 96 °C; $[\alpha]_D^{20}$ + 25.6 (*c* 1, CCl_4). Elemental analysis gives the expected values. This intermediate is only moderately stable and can be stored only for a short period of time.

N^{α}-*tert*-Butyloxycarbonyl-N^{π}-benzyloxymethyl-L-histidine Methyl Ester Hydrochloride

I + C_6H_5–CH_2O – CH_2Cl ⟶ $(CH_3)_3C$ – O – C(=O) – NH – CH – $COOCH_3$ (H_2C–imidazole, N – CH_2 – O – CH_2–C_6H_5, N+–H Cl^-)

II

A solution of the blocked amino acid ester (I, 30.2 g, 82 mmol) and freshly distilled benzyl chloromethyl ether [3] (18 ml, 130 mmol) in dichloromethane (200 ml) is allowed to stand at room temperature overnight. The solvent is removed in vacuo and ether (400 ml) is added to the residue. Next day, the crystalline material, which slowly separated, is collected on a filter, washed with ether and dried: 24 g (69%). It melts at 152 °C; $[\alpha]_D^{20}$ – 19.1 (*c* 1.0, MeOH). On elemental analysis the calculated values are found.

N^{α}-*tert*-Butyloxycarbonyl-N^{π}-benzyloxymethyl-L-histidine

II + NaOH ⟶ $(CH_3)_3C$ – O – C(=O) – NH – CH – COOH (H_2C–imidazole, N – CH_2 – O – CH_2–C_6H_5)

$C_{19}H_{25}N_3O_5$ (375.4)

To a solution of the ester hydrochlorid (II, 22 g, 52 mmol) in methanol (50 ml) a one molar solution of NaOH (120 ml) is added. After 15 minutes distilled water (one liter) is added and the pH of the solution is adjusted to 4.5 by the dropwise addition of N HCl. The mixture is extracted with chloroform (three times, 100 ml each time), the combined extracts dried over anhydrous sodium sulfate and the solvent removed in vacuo. The remaining oily residue is dissolved in ethyl acetate (50 ml) and the solvent evaporated in vacuo. The protected amino acid thus obtained (17 g, 87%) melts at 155 °C $[\alpha]_D^{20}$ + 6.9 (*c* 0.5, MeOH) and is analytically pure.

1. Brown T, Jones JH, Richards JD, J Chem Soc Perkin I 1982: 1553
2. Commercially available, as di-*tert*-butyl dicarbonate.
3. Commercially available.

3.6.3 N^{im}-Trityl-L-histidine [1]

Trityl-L-histidine [1]

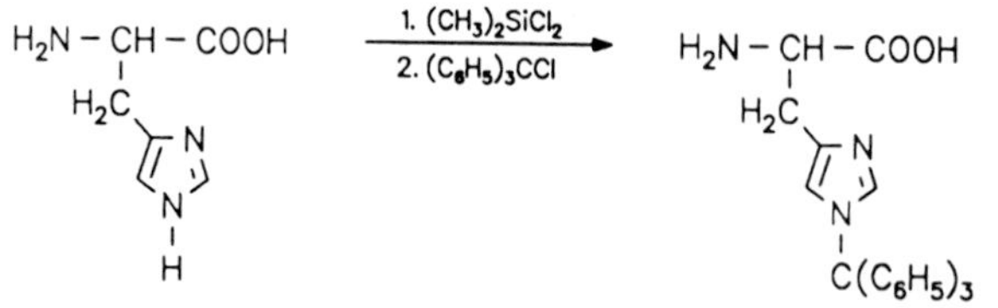

H$_2$N−CH−COOH, H$_2$C−(imidazolyl, N−H) → 1. $(CH_3)_2SiCl_2$; 2. $(C_6H_5)_3CCl$ → H$_2$N−CH−COOH, H$_2$C−(imidazolyl, N−$C(C_6H_5)_3$)

$C_{25}H_{23}N_3O_2$ (397.5)

Dichlorodimethylsilane (1.21 g, 10 mmol) is added to a stirred suspension of histidine (1.55 g, 10 mmol) in dichloromethane (15 ml) and the mixture is heated under a reflux condenser for 4 h. Triethylamine (2.8 ml, 20 mmol) is added and refluxing continued for 15 min. The reaction mixture is cooled to room temperature, stirred and treated with triethylamine (1.4 ml, 10 mmol) and then with a solution of trityl chloride (2.8 g, 10 mmol) in dichloromethane (10 ml). After two hours methanol is added and the solvents removed in vacuo. Water is added to the residue and the pH of the solution is adjusted to 8–8.5 by dropwise addition of triethylamine. The resulting slurry is thoroughly shaken with chloroform and the insoluble material collected on a filter by suction. It is washed with distilled water, then with ether, and dried: 3.85 g (97%), melting at 218–219 °C, it gives a single spot on chromatograms. For analysis a sample is recrystallized from a 1:1 mixture of tetrahydrofurane and water; this raises the m.p. to 220–222 °C; $[\alpha]_D^{25} -2.1$ (*c* 1, THF–H_2O, 1:1).

1. Barlos K, Papaioannou D, Theodoropoulos D (1982) J Org Chem 47: 1324

3.7 Blocking the Indole Nitrogen in Tryptophan

N^{in}-Formyl-L-tryptophan [1]

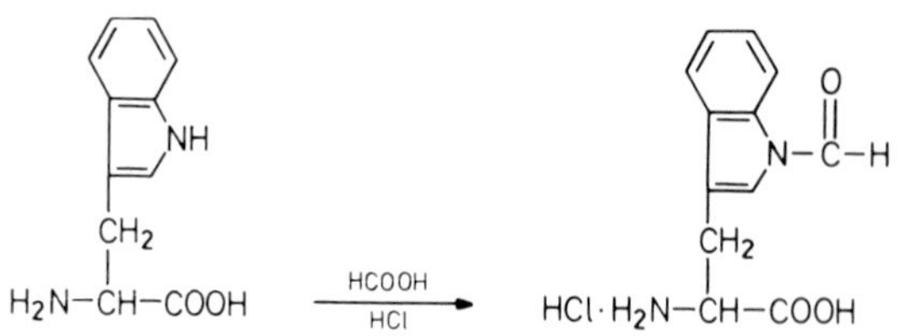

(indolyl NH)−CH$_2$−CH(NH$_2$)−COOH → HCOOH / HCl → (indolyl N−C(=O)−H)−CH$_2$, HCl·H$_2$N−CH−COOH

$C_{12}H_{12}N_2O_3 \cdot HCl$ (268.7)

Dry HCl gas is bubbled into a solution of L-tryptophan (20.4 g, 100 mmol) in formic acid (300 ml). At about hourly intervals samples [2] are taken from the mixture, diluted with water and their u.v. spectra recorded. The maximum at 278 nm, characteristic for tryptophan, gradually decreases and a new peak emerges at 298 nm. Formylation is complete when there is no more increase in the absorption at 298 nm. About three hours are necessary to reach this point. The solvent is removed in vacuo, ether is added to the remaining syrup and the crystals which form are collected on a filter. They are washed with ether and dried in air. The yield is quantitative (26.8 g). The product, N^{in}-formyl-L-tryptophan hydrochloride melts at 218–220 °C dec.; $[\alpha]_D^{23}$ −4.7° (c 2, H_2O). Correct values are obtained for C, H, and N on elemental analysis [3].

1. Ohno M, Tsukamoto S, Makisumi S, Izumiya N (1972) Bull Chem Soc Jpn 45: 2852
2. A 0.50 ml sample can be diluted with water to 10 ml and 1.0 ml of the dilute solution further diluted with water to 100 ml.
3. The formyl group is cleaved by nucleophiles. E.g. a solution of N^{α}-acetyl-N^{im}-formyl-tryptophan methyl ester is converted to a deformylated derivative (presumably N^{α}-acetyl-tryptophan hydrazide) by a 10% solution of hydrazine hydrate in dimethylformamide within 48 hours at room temperature.

3.8 Protection of Side Chain Carboxyl Groups

L-Aspartic Acid β-Benzyl Ester [1] (β-Benzyl L-aspartate)

$$H_2N\text{-}CH(CH_2COOH)\text{-}COOH + HOCH_2\text{-}C_6H_5 \xrightarrow[\text{2. pyridine}]{\text{1. } H_2SO_4} H_2N\text{-}CH(CH_2\text{-}CO\text{-}OCH_2\text{-}C_6H_5)\text{-}COOH$$

$C_{11}H_{13}NO_4$ (223.2)

Freshly distilled benzyl alcohol (100 ml) is added to a mixture of dry ether (100 ml) and concentrated sulfuric acid (10 ml). The ether is removed in vacuo and finely ground aspartic acid (13.4 g, 100 mmol) is added in small portions with stirring. The resulting solution is kept at room temperature for about a day, when it is diluted with 95% ethanol (200 ml) and neutralized by the dropwise addition of pyridine (50 ml) under vigorous stirring. The mixture is stored in the refrigerator overnight, the crystalline product collected on a filter and thoroughly washed by trituration on the filter with ether [2]. The ester is recrystallized from hot water containing a few drops of pyridine. The recovered material (about 9 g, 40%) is analytically pure. It melts at 218–220 °C; $[\alpha]_D^{25}$ +28° (c 1, N HCl).

1. Benoiton L (1962) Can J Chem 40: 570
2. A sinter-glass filter is recommended.

L-Glutamic Acid *γ-tert*-Butyl Ester [1]

$C_9H_{17}NO_4$ (203.2)

A solution of K_2CO_3 (16.8 g, 120 mmol) in water (100 ml) is vigorously stirred and cooled in an ice-water bath. L-Glutamic acid α-benzyl ester [2] (23.7 g, 100 mmol) is added. When almost all the ester is in solution benzyl chlorocarbonate (20.5 g = 17.1 ml, 120 mmol) is added in four about equal portions over a period of 30 min. The pH of the mixture is maintained around 8 by the periodic addition of a 10% solution of K_2CO_3 in water. (A total of 70 to 100 ml is needed.) Stirring is continued for 10 minutes longer. The pH is adjusted, if necessary, to about 8 and the mixture is extracted with ether (twice, 150 ml each time). The aqueous phase is acidified to Congo with 6 N HCl: the oil which separates solidifies on standing. The solid is collected on a filter, washed with water and dried in air. Recrystallization from ethanol-water or from tetrachloromethane affords pure benzyloxycarbonyl-L-glutamic acid α-benzyl ester [2] (26 to 31 g, 70–83%) melting at 95–96 °C; $[\alpha]_D^{24} -10.4°$ (*c* 1.7, AcOH).

A 50 mmol aliquot of benzyloxycarbonyl-L-glutamic acid α-benzyl ester (18.6 g) dissolved in dry dioxane (55 ml) is placed in a thick-walled glass vessel, cooled to −10 °C and mixed with liquid isobutylene (210 ml). After the addition of concentrated sulfuric acid (one ml) the flask is securely closed, wrapped in a towel and shaken at room temperature for about 20 hours. Water (250 ml) is added and the excess isobutylene is removed in vacuo. The residue is extracted with ether (twice, 150 ml each time). The ether extracts are pooled, extracted with a cold (0 °C) saturated solution of $NaHCO_3$ in water (four times, 100 ml each time), washed with ice water until the washes are neutral, dried over anhydrous $MgSO_4$ and evaporated to dryness in vacuo. The residue, a colorless oil, gradually crystallizes. The product, benzyloxycarbonyl-L-glutamic acid α-benzyl *γ-tert*-butyl ester [3], (about 13.5 g, 63%) melts at 40–44 °C [4].

A 10 mmol aliquot of the diester (4.3 g) is dissolved in a mixture of methanol (50 ml) and water (25 ml), a 10% Pd on charcoal catalyst (0.9 g) is added and the mixture hydrogenated [5] until no more CO_2 evolves [6]. The system is flushed with nitrogen, the catalyst is removed by filtration and the solvent by evaporation in vacuo. The residue is triturated with acetone and, after storage in the refrigerator overnight, filtered and washed with acetone. L-Glutamic acid γ-*tert*-butyl ester [7] (1.8 g, 89%) melts at 182 °C; $[\alpha]_D^{27}$ +9.8° (*c* 2, H_2O).

1. Zervas L, Hamalidis C (1965) J Amer Chem Soc 87: 99
2. Sachs H, Brand E (1953) J Amer Chem Soc 75: 4610; cf. also this volume, p. 32.
3. Schwyzer R, Kappeler H (1961) Helv Chim Acta 44: 1991
4. A sample can be recrystallized from ethyl acetate-hexane: m.p. 46–48 °C. For analysis it should be dried at room temperature in high vacuum.
5. Air should be displaced by nitrogen before hydrogen is allowed to enter the flask.
6. From time to time the escaping gas is tested with a half saturated aqueous solution of $Ba(OH)_2$.
7. Aspartic acid β-*tert*-butylester (m.p. 189–190 °C, $[\alpha]_D^{23}$ +8.5° (*c* 1, 90% AcOH)) is prepared in an analogous manner (Schwyzer R, Dietrich H (1961) Helv Chim Acta 44: 2003.

3.9 Protection of the Carboxamide Function

Introduction of the 4,4′-Dimethoxybenzhydryl Group [1]

$C_6H_5-CH_2O-CO-NH-CH(CH_2-CH_2-CONH_2)-COOH + CH_3O-C_6H_4-CH(OH)-C_6H_4-OCH_3 \xrightarrow{H_2SO_4}$

$C_6H_5-CH_2O-CO-NH-CH(CH_2-CH_2-CO-NH-CH(C_6H_4-OCH_3)_2)-COOH$

$C_{28}H_{30}N_2O_7$ (506.6)

A solution of benzyloxycarbonyl-L-glutamine (28.0 g, 100 mmol) and 4,4′-dimethoxybenzhydrol [2] (24.4 g, 100 mmol) in acetic acid (250 ml) is stirred at room temperature. Concentrated sulfuric acid (0.5 ml) is added and the mixture is allowed to stand overnight. The solution is poured into water (750 ml): the product separates as an oil which soon solidifies. The crystals are collected on a filter and are dissolved in ethyl acetate. The solution is washed with water, dried over anhydrous Na_2SO_4 and evaporated in vacuo to dryness. The residue is triturated with ether, filtered, washed with ether and dried. The protected amino acid weighs 45.8 g (90%) and melts at 117–120 °C; $[\alpha]_D^{22}$ −6.7° (*c* 2, dimethylformamide). On elemental analysis correct values are obtained for C, H and N [3–6].

1. König W, Geiger R (1970) Chem Ber 103: 2041
2. Commercially available.
3. If necessary, further purification can be achieved by reprecipitation from a solution in tetrahydrofuran with petroleum ether.

4. The same procedure can be applied for the preparation of benzyloxycarbonyl-4,4′-dimethoxybenzhydryl-L-asparagine except that the starting material, *N*-benzyloxycarbonyl-L-asparagine (27.0 g, 100 mmol) is dissolved in acetic acid (300 ml) with warming. The yield is 47.5 g (96%), m.p. 176–180 °C; $[\alpha]_D^{22}$ 2.4° (*c* 2, dimethylformamide). Catalytic hydrogenation in acetic acid selectively removes the benzyloxycarbonyl group.
5. The 4,4′-dimethoxybenzhydryl group is cleaved by trifluoroacetic acid, preferably in the presence of anisole [1]. Heating solutions of benzyloxycarbonyl-*N*-4,4′-dimethoxy-benzhydryl-L-glutaminyl peptides in a 9:1 mixture of trifluoroacetic acid and anisole to reflux for one and a half hour results in the formation of pyroglutamyl peptides (cf. König W, Geiger R (1972) Chem Ber 105: 2872
6. For a selectively removable protection of the carboxamide function by tritylation cf. Sieber P, Riniker B (1991) Tetrahedron Lett 32: 739

3.10 Blocking the Sulfhydryl Group in Cysteine

S-Benzyl-L-cysteine [1]

$H_2N-CH(COOH)-CH_2-S-S-CH_2-CH(COOH)-NH_2$ + 4 Na $\xrightarrow{\text{in } NH_3}$ 2 $H_2N-CH(CH_2-SNa)-COONa$ $\xrightarrow{ClCH_2-C_6H_5}$ $H_2N-CH(CH_2-S-CH_2-C_6H_5)-COONa$ $\xrightarrow{HCl}$ $H_2N-CH(CH_2-S-CH_2-C_6H_5)-COOH$

$C_{10}H_{13}NO_2S$ (211.3)

This reaction must be carried out in a well ventilated hood. Liquid ammonia [2] is condensed in a 3 liter round bottom flask provided with a powerful stirrer and with gas inlet and outlet tubes. The vessel is surrounded by a trichloroethylene [3] bath and a stream of ammonia is led through the flask to displace air from it. Dry ice is added to the bath and the stream of ammonia, regulated to maintain a slight positive pressure, is led into the flask [4] until it is about two thirds full with liquid ammonia. The NH_3 stream is reduced but the cooling bath is removed [5] before it is entirely cut off. Clean pieces of sodium metal [6] are added (about 10 g) followed by cystine until the blue color disappears. The alternating addition of the reactants is continued until the total amount of cystine (120 g, 0.5 mol) and sodium (48 g, 2.04 atoms) are used up. The blue color is discharged with a small amount of ammonium chloride and benzyl chloride (redistilled, 127 g = 116 ml, one mol) is added fairly rapidly. A thick mass of crystals separates. The ammonia is allowed to evaporate overnight. The flask with the solid residue is heated in a bath of 40–50 °C and evacuated with a water aspirator in order to remove residual ammonia. After about two hours, ice-water (approximately one liter) is added, the solution is treated with activated charcoal (about 5 g) and filtered. The filtrate is acidified to litmus with 6 N HCl [7], the precipitate is collected on a filter, washed with distilled water (1.5 liters) and with ethanol (0.5 liter). The product is dried in air until constant weight: 180 g (85%), m.p. 215–116 °C dec., $[\alpha]_D^{23}$ +24° (*c* 1, 1 N NaOH) [8].

1. Wood JL, du Vigneaud V (1939) J Biol Chem 130: 109
2. Anhydrous ammonia should be used. The cylinder, provided with a reducing valve, can be directly connected with the reaction vessel. The gas outlet tube should point toward the back of the hood, away from the operator. To facilitate the control of the pressure inside the flask, the excess ammonia should leave through a bubbler filled with silicone oil. A slight positive pressure throughout the collection of the liquid ammonia prevents the access of moist air.
3. Instead of trichloroethylene acetone or methanol can be used.
4. Reduction and benzylation are carried out at the boiling point of ammonia and the flask needs no protection other than the one provided by the escaping ammonia.
5. For the sake of safety, the cooling bath should be replaced by an empty vessel.
6. The metal, from which the crust of sodium hydroxide or sodium carbonate has been removed, is weighed and kept under hexane until used.
7. During the addition of hydrochloric acid the thick mass must be thoroughly mixed. Overacidification may result in the separation of the hydrochloride of *S*-benzyl-L-cysteine. If too much acid has been added the pH should be adjusted to about 6 with dilute ammonium hydroxide.
8. This material is pure enough for most practical purposes. If necessary, *S*-benzyl-L-cysteine can be recrystallized from a large volume of boiling water.

S-Ethylcarbamoyl-L-cysteine [1, 2]

$$Cl^- H_3\overset{+}{N}-CH(CH_2-SH)-COOH + O=C=N-C_2H_5 \longrightarrow H_2N-CH(CH_2-S-CO-NHC_2H_5)-COOH$$

$C_6H_{12}N_2O_3S$ (192.2)

L-Cysteine hydrochloride monohydrate is dried at 70 °C in vacuo overnight [3]. The dry material (15.8 g, 100 mmol) is dissolved in dimethylformamide (150 ml), the solution rapidly cooled to 0 °C and then treated, without delay, with ethyl isocyanate [4] (7.8 g = 8.7 ml, 110 mmol). The reaction mixture is stored at room temperature for 3 days. The volatile materials are removed in vacuo and the viscous residue is triturated with ether. The ether is decanted and the solid product dissolved in water (200 ml). The solution is extracted with ether (twice, 100 ml each time). The pH of the aqueous solution is adjusted to 6.5 and the solution is concentrated in vacuo until crystals appear (about 130 ml). After overnight storage in the refrigerator the crystals are collected on a filter, washed with ice-water and then with a 1:1 mixture of ether and 95% ethanol. A second crop is secured by concentration of the combined mother liquor and washings in vacuo to about half of their original volume. Again, the crystals are collected after overnight storage in the refrigerator. They are washed as the first crop. The total yield is 13.1 g (68%). *S*-Ethylcarbamoyl-L-cysteine melts at 219 °C dec.; $[\alpha]_D^{22}$ −91° (*c* 0.8, 95% AcOH); $[\alpha]_D^{22}$ −36.6°

(c 1.1, 6 N HCl). The product is chromatographically homogeneous and gives satisfactory values for C, H, N and S on elemental analysis.

1. Guttmann S (1966) Helv Chim Acta 49: 83
2. The *S*-ethylcarbamoyl (Ec) group is acid resistant but can be removed with nucleophiles. Accordingly, in the preparation of *N*-acyl derivatives, such as *N*-benzyloxycarbonyl-*S*-ethylcarbamoyl-L-cysteine an aqueous solution of $KHCO_3$ rather than NaOH is used [1] as acid binding agent.
3. Drying removes the water of crystallization from the starting material.
4. The ethyl isocyanate should be redistilled before used. It boils at 60 °C. Alkyl isocyanates are harmful materials and should be handled with care. The operations described above should be carried out in a well ventilated hood.

S-Acetamido-methyl-L-cysteine [1]

$$Cl^-H_3\overset{+}{N}-CH(CH_2-SH)-COOH + HO-CH_2-NH-CO-CH_3 \xrightarrow{CF_3COOH} Cl^-H_3\overset{+}{N}-CH(CH_2-S-CH_2-NH-CO-CH_3)-COOH$$

$C_6H_{13}N_2O_3SCl$ (228.7)

A mixture of L-cysteine hydrochloride (1.58 g, 10 mmol), *N*-hydroxymethyl-acetamide [2] (0.89 g, 10 mmol) and trifluoroacetic acid (10 ml) is stirred at room temperature for about 30 min. The trifluoroacetic acid is removed in vacuo, the residue dissolved in N HCl and evaporation repeated. The crude hydrochloride is crystallized from 2-propanol [3], washed with ether [3] and dried. The product (1.62 g, 71%) melts with decomposition at 155–157 °C. Recrystallization from 2-propanol [3] raises the m.p. to 166–168 °C dec. [4]; $[\alpha]_D$ $-33.2°$ (c 1, H_2O) [4].

1. Marbach P, Rudinger J (1974) Helv Chim Acta 57: 403
2. Acetamide (10 g) is hydroxymethylated by adding it to a solution of K_2CO_3 (1 g) in formaldehyde (12.3 g of a 41% solution). The mixture is heated on a steam bath for about 3 minutes and then allowed to stand at room temperature overnight. The solution is saturated with CO_2 and evaporated in vacuo, the residue is treated with anhydrous Na_2SO_4 and extracted with acetone. The acetone extracts are further dried with Na_2SO_4 and evaporated to dryness. The product, a colorless oil, solidifies on standing to a crystalline mass (m.p. 50–52 °C) which is quite hygroscopic. (Einhorn A (1905) J Liebigs Ann Chem 343: 265).
3. Peroxide free solvents should be used.
4. Veber DF, Milkowski JD, Varga SL, Denkewalter RG, Hirschmann R (1972) (J Amer Chem Soc 94: 5456) report a m.p. of 159–163 °C and $[\alpha]_D$ $-30.7°$ (c 1, H_2O).

S-Triphenyl-methyl-L-cysteine [1] (S-Trityl-L-cysteine)

$$Cl^-H_3\overset{+}{N}-CH(CH_2-SH)-COOH + HO-C(C_6H_5)_3 \xrightarrow[2.\ CH_3COONa]{1.\ BF_3\cdot(C_2H_5)_2O} H_2N-CH(CH_2-S-C(C_6H_5)_3)-COOH$$

$C_{22}H_{21}NO_2S$ (363.5)

A mixture of L-cysteine hydrochloride (15.8 g, 100 mmol) and glacial acetic acid (100 ml) is warmed in a 200 ml Erlenmeyer flask on a steam bath with occasional swirling at 60 °C. Triphenylmethanol [2] (26.0 g, 100 mmol) is added and the temperature is raised again to 60 °C before boron trifluoride etherate [3] (14 ml = 16.2 g, 114 mmol) is added in one portion. The solution is warmed on the steam bath to 80 °C for 30 min and then cooled to room temperature. About 45 min later the reaction mixture is transferred to a beaker with ethanol (150 ml), water (50 ml) and powdered anhydrous sodium acetate (30 g) are added. On dilution with water (400 ml) a gum separates which solidifies under cold water. The solid is disintegrated, thoroughly washed with water, then with acetone and finally with ether. It is dried over NaOH pellets, and P_2O_5. The product (30.8 g, 85%) melts, with decomposition at 181–182 °C. Recrystallization from dimethylformamide-water raises the m.p. to 183.5 °C, $[\alpha]_D^{24}$ +114° (*c* 0.8, 0.04 N ethanolic HCl) [4].

1. Hiskey RG, Adams JB, Jr (1965) J Org Chem 30: 1340
2. Triphenylcarbinol, trityl alcohol. It is commercially available.
3. Boron trifluoride etherate is volatile, inflammable and toxic. Thus, the operations should be carried out in a well ventilated hood.
4. Slightly lower m.p. and specific rotation were recorded in a process in which triphenylchloromethane (trityl chloride) was used for the alkylation of the sulfhydryl group. (Zervas L, Photaki I (1962) J Amer Chem Soc 84: 3887).

N-tert-Butyloxycarbonyl-*S*-(3-nitro-2-pyridinesulfenyl)-L-cysteine [1]

3-Nitro-2-pyridinesulfenyl Chloride

NO2, N, Cl + H2N, H2N, C=S → NO2, N, S–C, NH2, NH → (OH⁻) NO2, N, SH → ($K_3Fe(CN)_6$) NO2 O2N, N, S—S, N → (Cl_2) 2 NO2, N, SCl

$C_5H_3N_2O_2SCl$ (190.6)

2-Chloro-3-nitropyridine (15.8 g, 100 mmol) is added to a warm solution of thiourea (8 g, 105 mmol) in ethanol (160 ml), the mixture is boiled under a reflux condenser for 7 hours, then cooled to room temperature. The separated 3-nitro-2-pyridyl-pseudothiourea is collected on a filter, washed with ethanol and suspended in a solution of sodium carbonate (6.7 g) in water (200 ml). The mixture is cooled to 0 °C and a solution of sodium hydroxide (10 g) in water (200 ml) is added with vigorous stirring: the suspended precipitate gradually dissolves. A small amount of insoluble material is removed by filtration and the solution is acidified to pH 3 by the addition of concentrated hydrochloric acid. The precipitated 3-nitro-2-pyridine-thiol is collected on a filter, washed with water and dissolved in a dilute solution of sodium hydroxide (10 g) in water (200 ml). The solution is filtered and a solution of potassium ferricyanide (33 g) in water (200 ml) is added, with vigorous stirring at room temperature. The precipitated 3,3′-dinitro-2,2-dipyridyl disulfide is collected on a filter, washed with water and dried at 100 °C over P_2O_5 in vacuo. The product (11.5 g 74%) melts at 249–250 °C with decomposition.

A sample of the disulfide (8.7 g, 50 mmol), placed in a round bottom-flask provided with a drying tube ($CaCl_2$), is suspended in dichloromethane (175 ml). The solution is cooled in an ice-water bath while a stream of chlorine is introduced [2]: an almost clear solution forms. A small amount of tarry material is removed by filtration and the solvent is removed in vacuo at a bath temperature of 30 °C, leaving a residue, 3-nitro-2-pyridinesulfenyl chloride, which melts with decomposition at 217–222 °C. The yield (9.5 g) is quantitative.

N-tert-Butyloxycarbonyl-S-(3-nitro-2-pyridinesulfenyl)-L-cysteine [1]

3-nitro-2-pyridinesulfenyl chloride (NO_2, SCl) + $H_2N-CH(CH_2SH)-COOH$ $\xrightarrow{HCOOH}$ Cl^- $H_3\overset{+}{N}-CH(COOH)-H_2C-S-S$-(3-nitro-2-pyridyl)

$\xrightarrow{(CH_3)_3C-O-CO-S-\text{(4,6-dimethylpyrimidin-2-yl)}}$ $(CH_3)_3C-O-C(=O)-NH-CH(COOH)-H_2C-S-S$-(3-nitro-2-pyridyl)

$C_{13}H_{17}N_3O_6S_2$ (375.4)

L-Cysteine (1.21 g, 10 mmol) is dissolved in 90% formic acid (100 ml) and 3-nitro-2-pyridinesulfenyl chloride (2.1 g, 11 mmol) is added to the stirred solution. The reaction is allowed to proceed at room temperature for one hour. A small amount of insoluble material is removed by filtration and the crude disulfide hydrochloride is precipitated by the addition of ether (200 ml). The precipitate is collected on a filter, washed with ether and dried over P_2O_5 in vacuo. The product, a hemihydrate, (2.23 g, 70%) melts at 188–190 °C with decomposition; $[\alpha]_D^{22}$ +140° (*c* 1, MeOH). On elemental analysis the expected values are found. The salt (hemihydrate) (1.61 g, 5 mmol) and triethylamine (2.1 ml, 15 mmol) are dissolved in a mixture of methanol (100 ml) and dichloromethane (100 ml) and *tert*-butyl 4,6-dimethylpyrimidyl-2-mercaptyl carbonate (1.33 g, 5.5 mmol) is added. The mixture is allowed to stand at room temperature overnight. The solvent is removed in vacuo and the residue is suspended in a mixture of ethyl acetate (200 ml) and 5% aqueous citric acid solution (200 ml). The organic phase is extracted with a 5% sodium bicarbonate solution (twice, 200 ml each time) and the aqueous extracts are acidified to pH 3 by the addition of citric acid. The product is extracted with ethyl acetate

(600 ml in three portions), the extracts are pooled, washed with water, dried over anhydrous sodium sulfate and evaporated in vacuo. The residue is crystallized from ethyl acetate-ether. The purified product is dried in vacuo (1.29 g; 69%). It melts at 153–155 °C, with decomposition [3].

1. Bernatowicz MS, Matsueda R, Matsueda GR (1986) Int J Peptide Protein Res 28: 107
2. These operations require a well ventilated hood.
3. The *S*-Npys group is removed by mercaptanes and was used in the synthesis of mixed disulfides.

S-tert-Butyl-L-cysteine hydrochloride [1]

$$Cl^- \; H_3\overset{+}{N}-CH(CH_2SH)-COOH + (CH_3)_3C-OH \xrightarrow{H^+} Cl^- \; H_3\overset{+}{N}-CH(CH_2-S-C(CH_3)_3)-COOH$$

$C_7H_{16}NO_2SCl$ (213.7)

L-Cysteine hydrochloride (157.6 g, 1 mol) is added to a mixture of 2N HCl (450 ml) and *tert*-butanol (97 g, 1.3 mol) and the mixture is heated under a long reflux condenser for 12 hours [2]. It is then cooled and concentrated in vacuo. The crystals, which separate, are collected on a filter, washed with acetone and dried. The hydrochloride (209 g, 90%) melts at 198–200 °C. Chromatographically pure material melting at 204 °C, $[\alpha]_D^{20}$ +6.35° 2.5, N HCl) is obtained by recrystallization from 4N HCl.

Removal of the S-tert-Butyl Group

$$-NH-CH(CH_2-S-C(CH_3)_3)-C(=O)- \xrightarrow{o\text{-}O_2N\text{-}C_6H_4\text{-}Cl} -NH-CH(CH_2-S-S-C_6H_4\text{-}o\text{-}NO_2)-C(=O)-$$

$$\xrightarrow{HO-CH_2CH_2-SH} -NH-CH(CH_2SH)-C(=O)- + HO-CH_2-CH_2-S-S-C_6H_4\text{-}o\text{-}NO_2$$

An equimolar amount of *o*-nitrobenzenesulfenyl chloride [3] is added to the stirred solution of the *S-tert*-butyl-cysteine containing peptide in acetic acid. Stirring is continued until all the reagent is dissolved. After further two hours at room temperature the product is isolated by filtration or, if necessary, by evaporation of the solvent and precipitation with ether.

The *S-o*-nitrobenzenesulfenyl derivative is dissolved in methanol (10 ml per mmol) and treated with 2-mercaptoethanol (1.4 ml, 20 mmol per mmol of peptide). After three hours at room temperature water (60 ml per mmol) is added and the mixture thoroughly extracted with ethyl acetate. The free thiol group of the intermediate in the solution can be oxidized to the disulfide by air or other oxidizing agents.

1. Pastuszak JJ, Chimiak A (1981) J Org Chem 46: 1868
2. Some isobutene escapes. The procedure must be carried out in a well ventilated hood.
3. Also known as *o*-nitrophenylsulphenyl chloride.

3.11 Protection of the Thioether in Methionine

L-Methionine Sulfoxide [1]

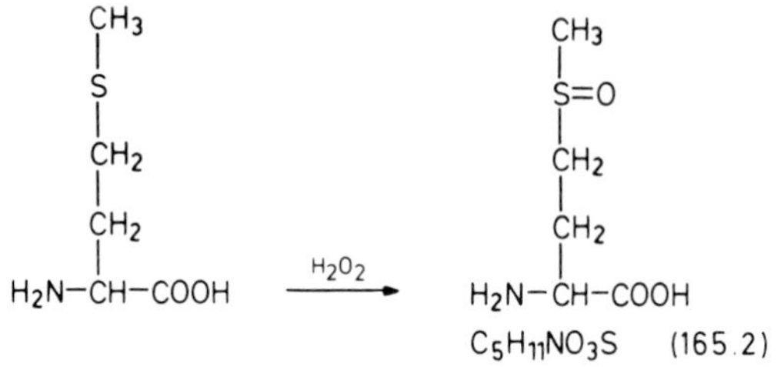

$$CH_3-S-CH_2-CH_2-CH(NH_2)-COOH \xrightarrow{H_2O_2} CH_3-S(=O)-CH_2-CH_2-CH(NH_2)-COOH$$

$C_5H_{11}NO_3S$ (165.2)

A suspension of L-methionine (3.0 g, 20 mmol) in distilled water (10 ml) [2] is surrounded by a water bath of room temperature and stirred vigorously. A solution of hydrogen peroxide (about 30%, 2.2 ml) is added in small portions over a period of 30 min. A clear solution forms. After a further hour at room temperature absolute ethanol (100 ml) is added. Two hours later the crystals are collected on a filter and washed with 95% ethanol (100 ml). The air-dry material (3.2 g, 97%) decomposes at about 253 °C; $[\alpha]_D^{22}$ +41.5° (*c* 2, N HCl) [3].

1. Iselin B (1961) Helv Chim Acta 44: 61. The original process in which oxidation with hydrogen peroxide was carried out in acetic acid, was modified by the present authors.
2. Preferably in a 125 ml Erlenmeyer flask.
3. The specific rotation originates from two chiral centers and can, therefore, vary from preparation to preparation (because of asymmetric induction the *d* and *l* sulfoxides are not produced in equal amounts).

***N-tert*-Butyloxycarbonyl-*S*-methyl-L-methionine *p*-Nitrophenyl Ester *p*-Toluenesulfonate [1]**

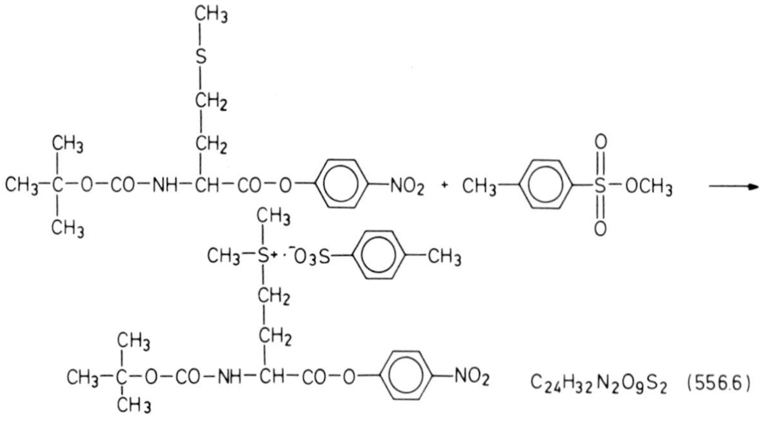

$$(CH_3)_3C-O-CO-NH-CH(CH_2-CH_2-S-CH_3)-CO-O-C_6H_4-NO_2 + CH_3-C_6H_4-SO_2-OCH_3 \longrightarrow$$

$$(CH_3)_3C-O-CO-NH-CH(CH_2-CH_2-S^+(CH_3)-CH_3 \cdot {}^-O_3S-C_6H_4-CH_3)-CO-O-C_6H_4-NO_2$$

$C_{24}H_{32}N_2O_9S_2$ (556.6)

Methyl *p*-toluenesulfonate [2] (18.6 g, 100 mmol) is added to a solution of *tert*-butyloxycarbonyl-L-methionine *p*-nitrophenyl ester [3] (3.7 g, 10 mmol) in ethyl acetate (17 ml) and the reaction mixture is stored at room temperature. After four days the first crop is collected on a filter, washed with a 1:1 mixture of ethyl acetate and ether and dried in vacuo: it weighs about 3 g. The combined filtrate and washings are concentrated in vacuo to a small volume and the solution is left to stand at room temperature for two days. The crystals are filtered, washed with ether and dried. The total of the two crops weighs about 5.2 g. The crude product is recrystallized from 95% ethanol which contains 1% acetic acid. The purified material (4.75 g, 85%) melts at 159–161 °C; $[\alpha]_D^{21}$ −30° (*c* 1, dimethylformamide containing 1% acetic acid). On thin layer chromatograms in the solvent system *n*-butanol-acetic acid-water (4:1:1) two spots can be detected by u.v. The fully protected amino acid active ester travels with an R_f value of 0.10 while the spot with R_f 0.60 corresponds to *p*-toluenesulfonic acid displaced from the salt during chromatography. The NMR spectrum, in AcOH-d_4, shows the expected signals; the two *S*-methyl groups appear as a singlet ∂ 3.06 downfield from TMS. The purified product gives correct C, H and N values on elemental analysis.

1. Bodanszky M, Bednarek MA (1982) Int J Peptide Protein Res 20: 408
2. Commercially available. It is a toxic material and should be handled with care. The operations should be carried out in a well ventilated hood.
3. Scoffone E, Rocchi R, Vidali G, Scatturin V, Marchiori I (1964) Gazz Chim Ital 943; C.A. 61: 13409b

III Activation and Coupling

1 The Acid Chloride Procedure

$$CH_3-C_6H_4-SO_2-NH-CH(CH(CH_3)CH_2CH_3)-COOH \xrightarrow{PCl_5} CH_3-C_6H_4-SO_2-NH-CH(CH(CH_3)CH_2CH_3)-COCl \xrightarrow{H_2NCH_2COOC_2H_5}$$

$$CH_3-C_6H_4-SO_2-NH-CH(CH(CH_3)CH_2CH_3)-CO-NH-CH_2-CO-OC_2H_5$$

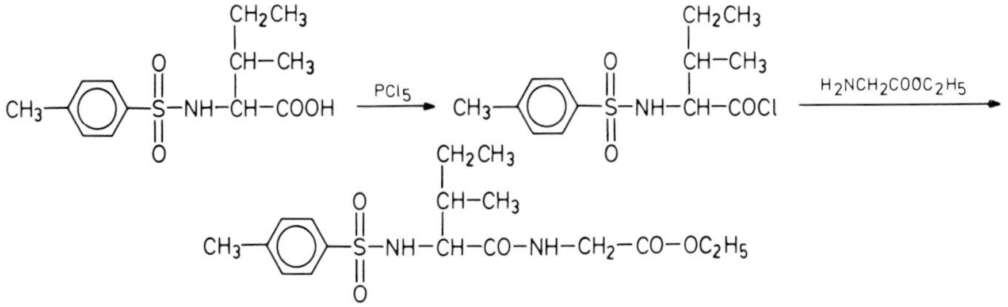

$C_{17}H_{26}N_2O_5S$ (370.5)

p-Toluenesulfonyl-L-isoleucyl-glycine Ethyl Ester [1]

p-Toluenesulfonyl-L-isoleucine [2] (2.9 g, 10 mmol) is dissolved in dry ether (15 ml), phosphorus pentachloride [3, 4] (2.2 g, 10.6 mmol) is added and the suspension is shaken until almost all the PCl_5 dissolves. The solution is decanted from the small amount of undissolved material and the ether evaporated in vacuo with a water aspirator. The phosphorus oxychloride remaining in the residue is removed by continuing the evaporation on an oil pump for about one hour. The acid chloride (about 3 g) is a colorless oil [5]. It is dissolved in dry ether and added, in several portions, to a suspension of glycine ethyl ester hydrochloride (2.5 g, 18 mmol) in dry ether (50 ml) containing triethylamine (3.5 ml = 2.52 g, 25 mmol). The mixture is allowed to stand at room temperature with occasional shaking for 12 hours. The precipitate is collected on a filter, washed with ether and then triturated and washed with water to extract triethylammonium chloride. The dry product (3.34 g, 90%) [6] melts at 159–160 °C. Recrystallization of a sample from ethanol raises the m.p. by 1 °C.

$$CH_3-C_6H_4-SO_2-NH-CH(CH(CH_3)CH_2CH_3)-COCl + H_2N-CH(CH_2CH_2CONH_2)-CO-NH-CH(CH_2CONH_2)-COOH \xrightarrow[2.\ HCl]{1.\ MgO}$$

$$CH_3-C_6H_4-SO_2-NH-CH(CH(CH_3)CH_2CH_3)-CO-NH-CH(CH_2CH_2CONH_2)-CO-NH-CH(CH_2CONH_2)-COOH$$

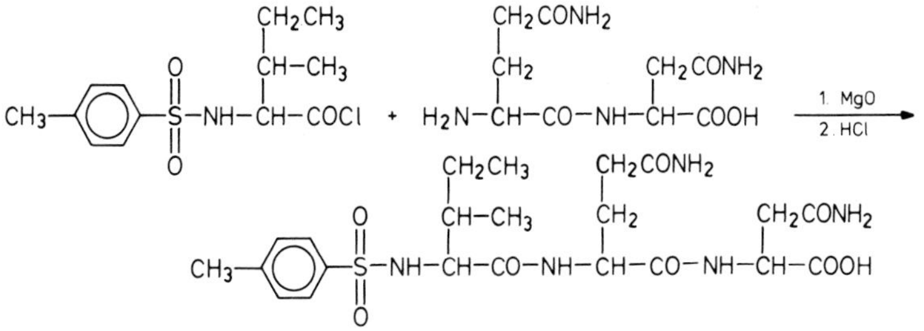

$C_{22}H_{33}N_5O_8S$ (527.6)

p-Toluenesulfonyl-L-isoleucyl-L-glutaminyl-L-asparagine [1]

Tosyl-L-isoleucine (2.9 g, 10 mmol) is converted to the acid chloride as described above. A solution of the acid chloride in dry dioxane (15 ml) is added, in small portions, over a period of about one hour, to a vigorously stirred (or shaken) ice cold suspension of L-glutaminyl-L-asparagine [7] (2.6 g, 10 mmol) and magnesium oxide (0.84 g, 21 mmol) in water (25 ml). The pH of the reaction mixture should be kept above 8 during the addition of the acid chloride [8]. The magnesium salt of the tosyltripeptide separates and the reaction mixture gradually turns into a semisolid mass. More water (12 ml) is added and the Mg salt is collected on a filter and washed with water. It is resuspended, without drying, in distilled water (300 ml) and the suspension acidified with 5% HCl to pH 3. The protected tripeptide acid is thoroughly washed on a filter with water and dried. The first crop weighs about 2.2 g. A second crop is secured by the acidification of the filtrates and washings from the Mg salt. The total amount of the crude product (m.p. 219–220 °C, dec.) is about 2.9 g (55%). It is dissolved in a 0.5 N solution of $KHCO_3$ in water and the solution is acidified with N HCl. The purified peptide derivative forms long needles, which melt with decomposition at 223 °C; $[\alpha]_D^{21}$ −31 ° (*c* 2, 0.5 N $KHCO_3$). On elemental analysis satisfactory values are obtained for C, H and N.

1. Katsoyannis PG, du Vigneaud V (1954) J Amer Chem Soc 76: 3114
2. Preparation of this tosylamino acid is described in this volume (p. 9).
3. The same acid chloride can be prepared by treating the tosylamino acid with thionyl chloride rather than with phosphorus pentachloride. Thionyl chloride, however, is not suitable for the conversion of benzyloxycarbonylamino acids into their chlorides because already at room temperature and even more on heating benzyloxycarbonylamino acid chlorides cyclize to *N*-carboxyanhydrides with the elimination of benzyl chloride. Treatment of *Z*-amino acids with PCl_5 should be carried out at 0 °C.
4. Phosphorus pentachloride is a dangerous material. Its contact with the skin and also with water should be carefully avoided.
5. The acid chloride prepared from tosyl-L-isoleucine with thionyl chloride was obtained in crystalline form by extraction of the distillation residue with boiling hexane. On cooling rectangular tables separated melting at 54–56 °C. The crystalline material gave satisfactory N and Cl values on elemental analysis (Bodanszky M, du Vigneaud V (1959)) J Amer Chem Soc 81: 2504
6. The yield can be further increased by extraction of the combined filtrate and washings with water, 0.5 N HCl, 0.5 N $KHCO_3$ and water and evaporation (after drying over anhydrous Na_2SO_4) to dryness. This second crop (about 0.25 g) after recrystallization from ethanol has the same m.p. as the first crop.
7. Swan JM, du Vigneaud V (1954) J Amer Chem Soc 76: 3110
8. On higher pH, however, most tosylamino acid chlorides decompose to *p*-toluenesulfonamide, carbon monoxide, an aldehyde and an (alkali) chloride. Therefore, it is advantageous to use MgO as acid binding agent, because its poor solubility in water prevents highly alkaline conditions.

2 The Azide Process

2.1 Peptide Hydrazides Through Hydrazinolysis of Peptide Esters

N-Benzyloxycarbonyl-S-benzyl-L-cysteinyl-L-seryl-L-histidine Hydrazide [1]

$C_6H_5-CH_2O-CO-NH-CH(CH_2-S-CH_2-C_6H_5)-CO-NH-CH(CH_2-OH)-CO-NH-CH(CH_2-\text{imidazolyl})-CO-OCH_3 + H_2NNH_2 \longrightarrow$

$C_6H_5-CH_2O-CO-NH-CH(CH_2-S-CH_2-C_6H_5)-CO-NH-CH(CH_2-OH)-CO-NH-CH(CH_2-\text{imidazolyl})-CO-NHNH_2 + CH_3OH$

$C_{27}H_{33}N_7O_6S$ (583.7)

Hydrazine hydrate (1.67 ml = 1.72 g, 34 mmol) is added to a solution of *N*-benzyloxycarbonyl-*S*-benzyl-L-cysteinyl-L-seryl-L-histidine methyl ester [1, 2] (5.84 g, 10 mmol) in methanol [3] (50 ml) and the reaction mixture is kept at room temperature for 3 days [4]. The separated hydrazide is collected by filtration and washed with methanol (20 ml), with a mixture of methanol (10 ml) and ether (10 ml) and with ether (30 ml). The crystalline peptide hydrazide, dried over P_2O_5 in vacuo, weighs 5.14 g (88%) and melts at 229 °C; $[\alpha]_D^{21}$ −7.6 (*c* 2, dimethylformamide).

1. Guttmann S, Boissonnas RA (1960): Helv Chim Acta 43: 200
2. Benzyl esters and ethyl esters react less readily with hydrazine than methyl esters.
3. Methanol is probably the solvent of choice for the hydrazinolysis of amino acid and peptide esters. If these are poorly soluble in methanol, hydrazinolysis can be attempted in suspension. Alternative solvents such as tetrahydrofurane, dioxane or their mixtures with methanol are also useful. Hydrazine, however, reacts with ethyl acetate and, albeit slowly, with dimethylformamide as well. In case of slow reactions additions of n-butanol might be helpful (Ferren RA, Miller JG, Day AR (1958) J Amer Chem Soc 79: 70. Hydrazinolysis of *S*-benzyl-L-cysteine containing peptides could be performed advantageously in 2-methoxyethanol at 30 °C; the reaction required 3 days (Maclaren JA, Savige WE, Swan JM (1958)): Austral J Chem 11: 345

4. It is common practice to carry out hydrazinolysis at elevated temperature, e.g. at the boiling point of the methanolic solution of the reactants. Because of the undesired side reactions which can occur under such conditions it is advisable to work at room temperature whenever possible.

2.2 Preparation of Peptide Hydrazides from Carboxylic Acids

tert-Butyloxycarbonyl-β-benzyl-L-aspartyl-L-leucyl-L-N^{ε}-benzyloxycarbonyl-lysine Hydrazide [1]

$(CH_3)_3C-O-CO-NH-CH(CH_2-CO-OCH_2C_6H_5)-CO-NH-CH(CH_2-CH(CH_3)_2)-CO-NH-CH((CH_2)_4-NH-CO-OCH_2C_6H_5)-COOH$ + H_2NNH_2 + $C_6H_{11}-N=C=N-C_6H_{11}$

$\xrightarrow{\text{HOBt}}$ $(CH_3)_3C-O-CO-NH-CH(CH_2-CO-OCH_2C_6H_5)-CO-NH-CH(CH_2-CH(CH_3)_2)-CO-NH-CH((CH_2)_4-NH-CO-OCH_2C_6H_5)-CO-NHNH_2$ + $C_6H_{11}NH-CO-NHC_6H_{11}$

$C_{36}H_{52}N_6O_9$ (712.8)

A sample of *tert*-butyloxycarbonyl-β-benzyl-L-aspartyl-L-leucyl-N^{ε}-benzyloxycarbonyl-L-lysine (7.0 g, 10 mmol) is dissolved in dimethylformamide (25 ml) and the stirred solution is cooled in an ice-water bath. Anhydrous hydrazine (0.38 g = 0.38 ml, 12 mmol) and 1-hydroxybenzotriazole monohydrate (3.4 g, 22 mmol) and finally dicyclohexylcarbodiimide (2.3 g, 11 mmol) are added. Stirring is continued at ice-water bath temperature for two hours and at room temperature overnight. After removal of the insoluble *N,N'*-dicyclohexylurea by filtration the solvent is evaporated in vacuo, the residue dissolved [2] in ethyl acetate, the solution washed with water, dried over anhydrous Na_2SO_4 and evaporated to dryness in vacuo. Recrystallization from methanol-ether affords the hydrazide in analytically pure form: 5.1 g (71%) melting at 133–137 °C; $[\alpha]_D^{25} - 19.4°$ (*c* 1, tetrahydrofuran).

1. Wang SS, Kulesha ID, Winter DP, Makofske R, Kutny R, Meienhofer J (1978): Int J Peptide Protein Research, 11: 297

2. If a hydrazide is insoluble in ethyl acetate the residue is triturated with water, filtered, thoroughly washed with water and, if possible, with 95% ethanol.

2.3 Protected Hydrazides

N-Benzyloxy-carbonyl-L-prolyl-glycyl-N′-*tert*-butyloxy-carbonyl-hydrazine [1]

$C_6H_5-CH_2O-CO-N-CH-CO-NH-CH_2-COOH + C_2H_5O-CO-Cl \xrightarrow{N(C_2H_5)_3}$

$C_6H_5-CH_2O-CO-N-CH-CO-NH-CH_2-CO-O-CO-OC_2H_5 \xrightarrow{H_2NNH-CO-O-C(CH_3)_3}$

$C_6H_5-CH_2O-CO-N-CH-CO-NH-CH_2-CO-NHNH-CO-O-C(CH_3)_3 + CO_2 + C_2H_5OH$

$C_{20}H_{28}N_4O_6$ (420.5)

A solution of benzyloxycarbonyl-L-prolyl-glycine [2] (30.6 g, 100 mmol) and triethylamine (10.1 g–14.0 ml, 100 mmol) in tetrahydrofurane (250 ml) is cooled to −10 °C and treated with ethyl chlorocarbonate [3] (10.9 g=9.6 ml, 100 mmol). Ten minutes later *tert*-butyl carbazate [4] (13.2 g, 100 mmol) is added and the mixture is allowed to warm up to room temperature and left to stand for about five hours. The solvent is removed in vacuo, the residue dissolved in ethyl acetate (300 ml) and water (150 ml), the organic layer washed with water (twice, 150 ml each time), with N NH_4OH (three times, 150 ml each time) and dried over anhydrous Na_2SO_4. The solvent is evaporated in vacuo, the residue dissolved in dry ether (75 ml) and kept in a refrigerator overnight. The crystals are collected on a filter, washed with dry ether (25 ml), with hexane (25 ml) and dried in vacuo: 19.5 g (46%), m.p. 119 °C: $[\alpha]_D^{21} -30°$ (*c* 2, dimethylformamide); −44.5° (*c* 2, methanol). Correct values are obtained on elemental analysis. Catalytic hydrogenation of the product in methanol affords *N*-L-prolyl-glycyl-*N′*-*tert*-butyloxycarbonylhydrazine in almost quantitative yield.

N-Benzyloxy-carbonyl-glycyl-glycyl-N′-*tert*-butyloxy-carbonyl-hydrazine [5]

$C_6H_5-CH_2O-CO-NH-CH_2-CO-NH-CH_2-COOH + H_2NNH-CO-O-C(CH_3)_3 \xrightarrow{C_6H_{11}-N=C=N-C_6H_{11}}$

$C_6H_5-CH_2O-CO-NH-CH_2-CO-NH-CH_2-CO-NHNH-CO-O-C(CH_3)_3 + C_6H_{11}-NH-CO-NH-C_6H_{11}$

$C_{17}H_{24}N_4O_6$ (380.4)

A suspension of benzyloxycarbonyl-glycyl-glycine [6] (2.66 g, 10 mmol) and *tert*-butyl carbazate [4] (1.32 g, 10 mmol) in methanol (70 ml) is cooled in an ice-water bath while dicyclohexylcarbodiimide [7] (2.06 g, 10 mmol) is added in several portions with stirring. Stirring is continued for one hour at 0 °C and at room temperature overnight. Once again the mixture is cooled with ice-water and glacial acetic acid (1.2 ml) is added. An hour later the undissolved,

N,N'-dicyclohexylurea is removed by filtration and the filtrate evaporated to dryness in vacuo. Tetrahydrofurane (about 10 ml) is added, the mixture cooled with ice-water and about an hour later filtered from the second portion of *N,N'*-dicyclohexylurea. The filtrate is diluted with ethyl acetate (80 ml), the solution washed with an ice-cold 10% solution of citric acid (50 ml) in water, with 0.5 N $KHCO_3$ (50 ml) and with water (twice, 50 ml each time). The solution is dried over anhydrous Na_2SO_4 and the solvent removed in vacuo. Crystallization from methanol-water affords the protected hydrazide (3.0 g, 79%) melting at 99–101 °C. The product is sufficiently pure for practical purposes [8]; hydrogenation in methanol in the presence of a Pd on charcoal catalyst removes the benzyloxycarbonyl group and permits the lengthening of the chain toward the *N*-terminus.

Glycine Benzyloxycarbonylhydrazide Trifluoroacetate [9]

$(CH_3)_3C\text{-}O\text{-}CO\text{-}NH\text{-}CH_2\text{-}COOH + Cl\text{-}C(=O)\text{-}OCH_2CH(CH_3)_2 \xrightarrow{CH_3\text{-}N(C_4H_8O)}$

$(CH_3)_3C\text{-}O\text{-}CO\text{-}NH\text{-}CH_2\text{-}C(=O)\text{-}O\text{-}C(=O)\text{-}O\text{-}CH_2CH(CH_3)_2 \xrightarrow{H_2NNH\text{-}CO\text{-}OCH_2\text{-}C_6H_5}$

$(CH_3)_3C\text{-}O\text{-}CO\text{-}NH\text{-}CH_2\text{-}CO\text{-}NHNH\text{-}CO\text{-}OCH_2\text{-}C_6H_5 \xrightarrow{CF_3COOH}$

$CF_3COO^- \cdot H_3\overset{+}{N}\text{-}CH_2\text{-}CO\text{-}NHNH\text{-}CO\text{-}OCH_2\text{-}C_6H_5 + CF_3CO\text{-}O\text{-}C(CH_3)_3$

$C_{12}H_{14}N_3O_5F_3$ (337.3)

A solution of *tert*-butyloxycarbonylglycine (17.5 g, 100 mmol) in ethyl acetate is cooled to −18 °C and treated with *N*-methylmorpholine (11.1 g = 12.1 ml, 110 mmol) and then with isobutyl chlorocarbonate [10] (13.7 g = 12.8 ml, 100 mmol). Four minutes later benzyl carbazate [11] (18.3 g, 110 mmol) is added and stirring is continued for 15 minutes at −18 °C and at room temperature for 45 minutes. The mixture is evaporated in vacuo, the residue taken up in ethyl acetate (200 ml) and water (100 ml) and the organic layer washed with an ice cold 10% solution of citric acid in water (100 ml), 0.5 N $KHCO_3$ (100 ml) and water (twice, 100 ml each time) and dried over an hydrous Na_2SO_4. The solvent is removed in vacuo, and the residue dissolved in 90% (aqueous) trifluoroacetic acid (80 ml). Fifteen minutes later the solution is evaporated to dryness in vacuo, the residue precipitated from ether with hexane and recrystallized from methanol-ether. The trifluoroacetate salt (29.3 g, 87%) melts at 176–177 °C and is analytically pure.

1. Boissonnas RA, Guttmann S, Jaquenoud PA (1960) Helv Chim Acta 43: 1349
2. Rydon HN, Smith PWG (1956) J Chem Soc 3642

3. Commercially available as ethyl chloroformate.
4. Commercially available. Its preparation is described in this volume: cf. p. 95.
5. Schwyzer R, Tun-Kyi A (1962) Helv Chim Acta 45: 859
6. Bergmann M, Zervas L (1932) Ber Dtsch Chem Ges 65: 1192
7. Dicyclohexylcarbodiimide is a known allergen. It has to be handled with care.
8. Chromatography on a silica gel column with chloroform as eluent followed by a second crystallization from methanol-water yields an analytical sample melting at 100–101 °C.
9. Storey HT, Beacham J Cernosek SF, Finn FM, Yanaihara C, Hofmann K (1972) J Amer Chem Soc 94: 6170
10. Commercially available as isobutyl chloroformate.
11. Commercially available (Eastman) as carbobenzoxyhydrazide.

2.4 Conversion of Hydrazides to Azides and Coupling with Azides

2.4.1 Generation of the Azide with the Aid of Sodium Nitrite and Coupling (Without Isolation of the Azide)

Benzyloxy-carbonyl-L-seryl-L-tyrosyl-L-seryl-L-methionyl-γ-benzyl-L-glutamic Acid [1]

C_6H_5–CH_2O–CO–NH–CH(CH_2OH)–CO–NH–CH(CH_2–C_6H_4–OH)–CO–NH–CH(CH_2OH)–CO–NHNH_2 + HONO ⟶

C_6H_5–CH_2O–CO–NH–CH(CH_2OH)–CO–NH–CH(CH_2–C_6H_4–OH)–CO–NH–CH(CH_2OH)–CO–N=$\overset{+}{N}$=$\overset{-}{N}$

$\xrightarrow{H_2N\text{-}CH(CH_2CH_2SCH_3)\text{-}CO\text{-}NH\text{-}CH(CH_2CH_2CO\text{-}OCH_2C_6H_5)\text{-}COOH}$

C_6H_5–CH_2O–CO–NH–CH(CH_2OH)–CO–NH–CH(CH_2–C_6H_4–OH)–CO–NH–CH(CH_2OH)–CO–NH–CH(CH_2–CH_2–S–CH_3)–CO–NH–CH(CH_2–CH_2–CO–OCH_2–C_6H_5)–COOH

$C_{40}H_{49}N_5O_{13}S$ (839.9)

A solution of benzyloxycarbonyl-L-seryl-L-tyrosyl-L-serine hydrazide [1] (5.04 g, 10 mmol) in dimethylformamide (50 ml) is cooled to −5 °C while 4 N HCl (10 ml) is added with stirring. This is followed by the addition of a precooled (0 °C) mixture of dimethylformamide (50 ml) and a 5 N solution of $NaNO_2$ in water (2.5 ml). Five minutes later the excess HCl is neutralized with triethylamine [2] (2.78 g = 3.8 ml, 27.5 mmol) and the solution is dried, briefly, over anhydrous Na_2SO_4. A second solution containing L-methionyl-γ-benzyl-L-glutamic acid [1] (5.53 g, 15 mmol) and triethylamine [2] (3.03 g = 4.2 ml, 30 mmol) in dimethylformamide (50 ml) is prepared and cooled in an ice water bath. The solution of the azide is filtered directly into the solution of the dipeptide derivative and the mixture is stirred at 0 °C for 36 hours. The solvent is removed in vacuo and the residue dissolved in 0.5 N NH_4OH (75 ml). Soon a solid starts to appear which is separated from the mother liquor by centrifugation [3]. This material, the ammonium salt of the target compound, is dissolved in methanol (1250 ml) and the solution passed through a column of the ion-exchange resin Amberlite IRA-120 and then through Amberlite IRA-45 (in acetate cycle). Evaporation to dryness, dissolution of the residue in boiling methanol (25 ml) concentration to about half of the original volume and dilution with ether (125 ml) yields the title compound (4.22 g, 50%) with a m.p. of 170 °C. An analytical sample melting at 173 °C can be obtained by two recrystallizations from methanol-ethyl acetate: $[\alpha]_D^{23} -23°$ (*c* 1, methanol) [4].

1. Guttmann S, Boissonnas RA (1958) Helv Chim Acta 41: 1852
2. The usually minor recemization which occurs in azide couplings is further diminished if triethylamine is replaced by an equivalent amount of diisopropylethylamine. Reduction of the base concentration is similarly helpful.
3. The precipitate can be filtered only with difficulty and during the prolonged operation some attack of ammonia on the benzyl ester group can take place.
4. Coupling through azides is accompanied by Curtius rearrangement. Since the resulting isocyanates yield urea derivatives, which are not readily separated from the desired product, this side reaction must be kept at a minimum. Hence, the azides should not be stored but used without delay and it is preferable to prepare the solution of the amino-component (in this case methionyl-γ-benzyl-glutamate) before or during the conversion of the hydrazide to the azide. The solutions of the two components are cooled before coupling and the coupling is carried out in the cold.

2.4.2 Conversion of a Hydrazide to the Azide with the Aid of Butyl Nitrite [1] and Acylation with the Azide "In Situ" [2]

Benzyloxy-carbonyl-L-tyrosyl-L-isoleucyl-L-glutaminyl-L-asparaginyl-*S*-benzyl-L-cysteinyl-L-prolyl-L-leucyl-glycinamide [3]

Z–Tyr–Ile–Gln–Asn–$NHNH_2$ + C_4H_9–O–NO $\xrightarrow{HCl}$ Z–Tyr–Ile–Gln–Asn–N_3

$\xrightarrow{\text{H-Cys(Bzl)-Pro-Leu-Gly-NH}_2}$ Z–Tyr–Ile–Gln–Asn–Cys(Bzl)–Pro–Leu–Gly–NH_2

$C_{55}H_{75}N_{11}O_{13}S \cdot 1/2H_2O$ (1139.3)

Benzyloxycarbonyl-L-tyrosyl-L-isoleucyl-L-glutaminyl-L-asparagine hydrazide [3] (0.69 g, 1 mmol) is dissolved in a mixture of dimethylformamide (25 ml) and 6.3 M HCl in tetrahydrofurane (1 ml). The mixture is cooled to −30 °C and freshly distilled *n*-butyl nitrite [4] (0.21 g = 0.24 ml, 2 mmol) is added followed, four minutes later, by a precooled (−30 °C) solution of *S*-benzyl-L-cysteinyl-L-prolyl-L-leucyl-glycine amide [5] (0.48 g, 1 mmol) in dimethylformamide (10 ml). The pH is adjusted to 8–9 (moist indicator paper) with *N*-ethylpiperidine (about 1 ml) and the reaction mixture is stored at 0 °C overnight. The solvent is removed in vacuo and the residue triturated with N HCl. The solid is collected on a filter, washed with N HCl, water, warm 0.5 N $KHCO_3$ and water. The dried product weighs 0.90 g (79%) and melts at 222–224 °C (hot stage m.p.): $[\alpha]_D -34°$ (*c* 0.2, dimethylformamide). The octapeptide derivative is chromatographically homogeneous. For analysis it is recrystallized from aqueous acetic acid; the melting point remains unchanged. The C, H and N values agree well with the ones calculated for a hemihydrate. Hydrolysis of a partially deblocked (HBr/AcOH) sample gives, on amino acid analysis, the expected ratios of the constituents.

1. Honzl J, Rudinger J (1961) Collect Czechoslov Chem Commun 26: 2333
2. Guttmann S, Boissonnas RA (1958) Helv Chim Acta 41: 1852
3. Jost K (1966) Collect Czechoslov Chem Commun 31: 2784
4. Instead of *n*-butyl nitrite, *tert*-butyl nitrite can be used with similarly good results; cf. Hofmann K, Haas W, Smithers MJ, Wells RD, Wolman Y, Yanaihara N, Zanetti G (1965) J Amer Chem Soc 87: 620
5. Zaoral M, Rudinger J (1955) Collect Czechoslov Chem Comm 20: 1183

2.4.3 Conversion of Protected Amino Acid or Peptide Hydrazides to the Azides Followed by Acylation with the Isolated Azide [1]

Azide coupling

$$Z{-}NH{-}CHR{-}CO{-}NHNH_2 + HONO \longrightarrow Z{-}NH{-}CHR{-}CO{-}N{=}\overset{+}{N}{=}\overset{-}{N}$$

$$Z{-}NH{-}CHR{-}CO{-}N_3 + H_2N{-}CHR'{-}COOR'' \longrightarrow Z{-}NH{-}CHR{-}CO{-}NH{-}CHR'{-}COOR'' + HN_3$$

The protected amino acid (or peptide) hydrazide (10 mmol) is dissolved [2] in a mixture of glacial acetic acid (12 ml), 5 N HCl (5 ml) and water (50 ml) and the solution is cooled to −5 °C. A solution of $NaNO_2$ (0.73 g, 10.6 mmol) in water (5 ml) is added. This causes the separation of the azide as an oil or as a solid. The azide is extracted into cold ether (60 ml), the extract cooled in an ice-water bath while washed with ice-water (40 ml). After brief drying over anhydrous Na_2SO_4, the ethereal solution of the azide is added to the solution of the amino-component [3] in chloroform (20 ml). The reaction mixture is allowed to stand at room temperature [4] for about 20 hours. The solution is extracted with 0.5 N HCl (40 ml), water (40 ml), 0.5 N $KHCO_3$ (40 ml) and water (40 ml), dried over Na_2SO_4 and concentration in vacuo. The product is crystallized from a suitable solvent [5].

1. From the numerous descriptions of azide couplings a procedure (Erlanger BF, Brand E (1951) J Amer Chem Soc 73: 2508) used for the preparation of benzyloxycarbonyldipeptide benzyl esters (e.g., benzyloxycarbonyl-L-alanyl-L-alanine benzyl ester) was selected. This method is applicable when the solubility of the carboxyl and amino components is sufficiently high in relatively non-polar solvents, which are not miscible with water. For the preparation

of larger peptides usually more polar solvents, mostly dimethylformamide, must be applied and, therefore, it is advisable to follow one of the procedures in which the azide is not isolated but used "in situ". In our view isolation of intermediates is not an unnecessary evil, but rather an advantageous operation. In the case of the azide procedure, however, the advantages of isolation of the reactive intermediate are offset by its unstable character. Thus, Curtius rearrangement of acid azides can be slowed down by working at low temperature but cannot be completely stopped. Hence, it seems to be better to forgo the often time-consuming isolation and to proceed after the formation of the azide direct to the acylation step.

2. With benzyloxycarbonyl-derivatives gentle heating, if necessary, is permissible, but not with amino acids or peptide hydrazides which are protected by highly acid labile blocking groups.
3. Amino acid esters are best used as such rather than a mixture of their salts with tertiary amines.
4. Lower temperature (e.g., 0 °C) is preferable from the point of view of by-product formation through Curtius rearrangement. Of course, a longer reaction time might be necessary. During the reaction HN_3 forms and can, in part, escape. It is a toxic substance.
5. With numerous small peptides crystallization can be effected by dissolution in ethyl acetate followed by dilution with hexane.

3 Symmetrical Anhydrides

tert-Butyloxycarbonyl-L-phenylalanine Anhydride [1, 2]

$$2\,(CH_3)_3C\text{-}O\text{-}CO\text{-}NH\text{-}CH(CH_2C_6H_5)\text{-}COOH + CH_3CH_2\text{-}N{=}C{=}N\text{-}CH_2CH_2CH_2N(CH_3)_2\cdot HCl \longrightarrow$$

$$(CH_3)_3C\text{-}O\text{-}CO\text{-}NH\text{-}CH(CH_2C_6H_5)\text{-}CO\text{-}O\text{-}CO\text{-}CH(CH_2C_6H_5)\text{-}NH\text{-}CO\text{-}O\text{-}C(CH_3)_3$$

$C_{28}H_{36}N_2O_7$ (512.6) + $CH_3CH_2NH\text{-}CO\text{-}NHCH_2CH_2CH_2N(CH_3)_2\cdot HCl$

A solution of *tert*-butyloxycarbonyl-L-phenylalanine (5.30 g, 20 mmol) in dichloromethane (200 ml) is placed into a 500 ml round bottom flask provided with a drying tube filled with cotton. The solution is cooled in an ice water bath and stirred vigorously with a magnetic stirrer. The water soluble carbodiimide [3] *N*-ethyl-*N*'-3-dimethylaminopropylcarbodiimide hydrochloride (1.92 g, 10 mmol) is added and stirring is continued at ice bath temperature for 2 hours. The solvent is evaporated in vacuo at a bath temperature of 0 °C and the residue dissolved in a mixture of ethyl acetate (200 ml) and water (100 ml) to which a few pieces of ice have been added. The organic layer is washed twice with ice water (100 ml each time), then with 0.5 N $KHCO_3$ and finally with saturated sodium chloride solution, always with pieces of ice added. The solution is dried over $MgSO_4$ while cooled in an ice-water bath, filtered and the solvent removed in vacuo at a bath temperature of 0 °C. The residue is crystalline or crystallizes on storage in the cold (−5 °C). The anhydride is washed with a 20:1 mixture of hexane and ether and dried in vacuo. The product, 3.9 g (76%), melts at 95–97 °C; $[\alpha]_D^{23} - 32.8°$ (*c* 2, $CHCl_3$) [4]. In the i.r. spectrum, in addition to the urethane carbonyl band (1710 cm^{-1}) two carbonyl bands, characteristic for anhydrides, appear at 1750 cm^{-1} and 1830 cm^{-1}.

1. Chen FMF, Kuroda K, Benoiton NL (1978) Synthesis, 928
2. The preparation of several additional *tert*-butyloxycarbonyl- and benzyloxycarbonyl-amino acid anhydrides is described in ref. 1. Symmetrical anhydrides of benzyloxycarbonylamino

acids were obtained also through the disproportionation of mixed anhydrides. (Wieland T, Kern W, Sehring R (1950) J Liebigs Ann Chem 569: 117; Wieland T, Flor F, Birr C, ibid. 1973: 1595

3. The expected formation of symmetrical anhydrides in the reaction of carbodiimides with protected amino acids (Khorana G (1955) Chem Ind 1087) was soon found to be useful in practical synthesis (Muramatsu I, Hagitani A (1959) Nippon Kagaku Zasshi 80: 1497; (1961) Chem Abstr 55: 6394)
4. Symmetrical anhydrides of benzyloxycarbonylamino acids could be obtained, with the help of dicyclohexylcarbodiimide, in crystalline form (Schüssler H, Zahn H (1962) Chem Ber 95: 1076). Benzyloxycarbonyl-glycine (2.1 g, 10 mmol) in acetonitrile (60 ml) was treated with a solution of dicyclohexylcarbodiimide (1.0 g, ca 5 mmol) dissolved in acetonitrile (10 ml). After 17 hours the precipitated urea derivative was removed by filtration and the solvent by evaporation in vacuo. The residue was crystallized from dry benzene to afford the anhydride of benzyloxycarbonylglycine in 75% yield, with m.p. 108–114 °C. The method cited in ref. 1 and described in detail above produced this anhydride with higher m.p. (118–119 °C) in 80% yield. Similarly higher yield and higher m.p. was noted when a water soluble carbodiimide rather than dicyclohexylcarbodiimide was used for the preparation of benzyloxycarbonyl-L-valine anhydride (cf. ref. 1).

Symmetrical anhydrides of N^{α}-9-fluorenylmethyloxycarbonylamino acids (Fmoc-amino acids) were prepared (Heimer EP, Chang CD, Lambros T, Meienhofer J (1981) Int J Peptide Protein Res 18: 237) through the reaction of the protected amino acids with a water soluble carbodiimide. A solution of the Fmoc-amino acid (20 mmol) in ethyl acetate (250 ml) is cooled in an ice-water bath while *N*-ethyl-*N'*-3-dimethylaminopropylcarbodiimide hydrochloride (2.11 g, 11 mmol) is added with stirring. Stirring is continued at 0 °C for one hour and at room temperature for an additional hour. The mixture is poured into ice water (250 ml), more ethyl acetate is added to dissolve the product and the organic phase is washed with 0.1 N HCl, water, 0.1 N $KHCO_3$, water, dried over $MgSO_4$ and evaporated to dryness in vacuo. Crystallization from ethyl-acetate-hexane affords the pure symmetrical anhydrides. Yields range from 50 to 86%. The Fmoc-amino acid anhydrides have the characteristic bands (1830 and 1750 cm^{-1}) in their i.r. spectra.

For extended periods, symmetrical anhydrides of *tert*-butyloxycarbonyl-amino acids should be stored at or below −20 °C, under anhydrous conditions (Yamashiro D (1987) Int J Peptide Protein Res 30: 9).

4 Mixed Anhydrides

4.1 The Isovaleric Acid Mixed Anhydride Method [1]

Benzyloxy-carbonyl-L-leucyl-glycine Ethyl Ester [1]

$$C_6H_5-CH_2O-CO-NH-CH(CH_2-CH(CH_3)_2)-COO^- \overset{+}{H}N(C_2H_5)_3 + (CH_3)_2CHCH_2CO-Cl \longrightarrow$$

$$C_6H_5-CH_2O-CO-NH-CH(CH_2-CH(CH_3)_2)-CO-O-CO-CH_2-CH(CH_3)_2 + (C_2H_5)_3\overset{+}{N}H \cdot Cl^- \xrightarrow[N(C_2H_5)_3]{Cl^- \cdot H_3\overset{+}{N}CH_2CO-OC_2H_5}$$

$$C_6H_5-CH_2O-CO-NH-CH(CH_2-CH(CH_3)_2)-CO-NH-CH_2-CO-OC_2H_5 + (C_2H_5)_3\overset{+}{N}H \cdot Cl^- + (CH_3)_2CHCH_2COOH$$

$C_{18}H_{26}N_2O_5$ (350.4)

A solution of benzyloxycarbonyl-L-leucine (2.65 g, 10 mmol) and triethylamine (1.01 g = 1.4 ml, 10 mmol) in a mixture of toluene (12.5 ml) and chloroform (12.5 ml) is cooled to $-5\,°C$. Isovaleryl chloride (1.205 g = 1.27 ml, 10 mmol) is added and the mixture allowed to stand at $-5\,°C$ for one and a half hours for the formation of the mixed anhydride. At that time a solution of glycine ethyl ester hydrochloride (1.4 g, 10 mmol) and triethylamine (1.4 ml, 10 mmol) in chloroform (25 ml), cooled to $-5\,°C$, is added and the reaction mixture is set aside in a refrigerator at about 5–10 °C. Next day the solution is washed with equal volumes of water, 0.5 M $KHCO_3$ solution, and water again. It is dried over $MgSO_4$, filtered from the drying agent and concentrated in vacuo to about 15 ml. On dilution with hexane (75 ml) the product separates in crystalline form. It is collected on a filter, washed with hexane and dried in air. The crude dipeptide derivative is dissolved in boiling 95% ethanol on a steam bath and diluted with warm distilled water until turbid. On cooling, crystals separate. After several hours at room temperature the product is collected on a filter, washed with a small volume of 50% ethanol, dried in air and finally in vacuo over P_2O_5. The protected dipeptide ester weighs 2.45 g (70%) [2] and melts at 105–106 °C; $[\alpha]_D^{24}$ −25.6° (*c* 5, ethanol).

1. Vaughan JR, Jr, Osato RL (1951) J Amer Chem Soc 73: 5553
2. The same method used in the coupling of benzyloxycarbonyl-L-proline to L-leucyl-glycine ethyl ester yielded benzyloxycarbonyl-L-prolyl-L-leucyl-glycine ethyl ester with m.p. 145–146 °C. A single recrystallization from aqueous ethanol afforded the protected tripeptide derivative in analytically pure form. (Ressler C, du Vigneaud V (1954) J Amer Chem Soc 76: 3107)

4.2 Trimethylacetic Acid (Pivalic Acid) Mixed Anhydrides [1, 2]

Benzyloxycarbonyl-α-methylalanyl-α-methylalanine Methyl Ester [1]

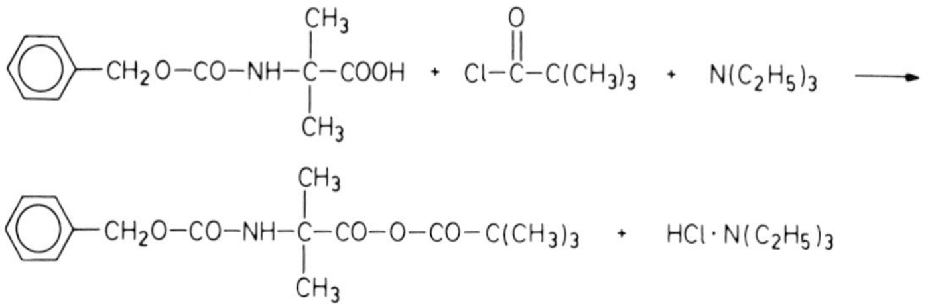

$C_6H_5-CH_2O-CO-NH-C(CH_3)_2-COOH + Cl-CO-C(CH_3)_3 + N(C_2H_5)_3 \longrightarrow$

$C_6H_5-CH_2O-CO-NH-C(CH_3)_2-CO-O-CO-C(CH_3)_3 + HCl \cdot N(C_2H_5)_3$

$C_{17}H_{23}NO_5$ (321.4)

$C_6H_5-CH_2O-CO-NH-C(CH_3)_2-CO-O-CO-C(CH_3)_3 + H_2N-C(CH_3)_2-CO-O-CH_3 \longrightarrow$

$C_6H_5-CH_2O-CO-NH-C(CH_3)_2-CO-NH-C(CH_3)_2-CO-OCH_3$

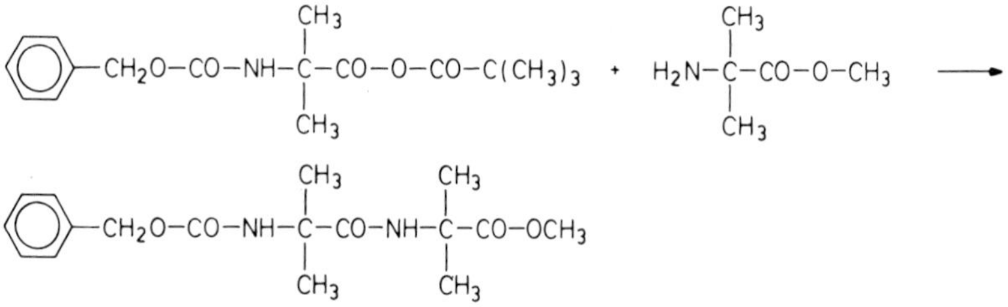

$C_{17}H_{24}N_2O_5$ (336.4)

Benzyloxycarbonyl-α-methylalanine (2.37 g, 10 mmol) and triethylamine (1.01 g = 1.40 ml, 10 mmol) are added to dry toluene (5 ml). The solution is cooled to −5 °C and treated with trimethylacetyl chloride (pivaloyl chloride, 1.21 g = 1.23 ml, 10 mmol). The mixture is stirred at −5 °C for two hours then at room temperature for one hour. The precipitated triethylammonium chloride is removed by filtration and the filtrate evaporated to dryness in vacuo. The residue, the mixed anhydride (3.17 g, 99%), melts at 81–83 °C; its i.r. spectrum shows carbonyl bands at 1805 and 1736 cm^{-1}, characteristic for anhydrides of carboxylic acids. Satisfactory values are obtained for C, H and N on elemental analysis.

The anhydride (3.21 g, 10 mmol) and α-methylalanine methyl ester (1.29 g, 11 mmol) are added to dry toluene (50 ml), the solution is heated at 60 °C for three hours and allowed to stand at room temperature overnight. It is diluted with ethyl acetate (50 ml), washed with 0.5 N $KHCO_3$ (50 ml), water (50 ml), 0.5 N HCl (50 ml) and water (50 ml), dried over anhydrous Na_2SO_4 and evaporated to dryness in vacuo. Crystallization of the residue from ether-hexane yields 2.95 g (88%) of the dipeptide derivative melting at 107–109 °C. Recrystallization from the same solvents can raise the m.p. to 109–111 °C.

N^α-Benzyloxy-carbonyl-N^ε-*p*-toluene-sulfonyl-L-lysylglycine Ethyl Ester [2]

$C_{25}H_{33}N_3O_7S$ (519.6)

N^α-Benzyloxycarbonyl-N^ε-tosyl-lysine [3] (4.34 g, 10 mmol) is dissolved in chloroform (20 ml) by the addition of pyridine (0.79 g = 0.80 ml, 10 mmol). The solution is cooled to −3 °C and treated with trimethylacetyl chloride (pivaloyl chloride, 1.21 g = 1.23 ml, 10 mmol). After about 10 minutes at 0 to −3 °C the mixture is cooled to −10 °C and glycine ethyl ester [4] (1.03 g, 10 mmol) is added. The reaction mixture is allowed to warm up and is kept at room temperature for about 30 min. The solvent is removed in vacuo, the residue shaken with a mixture of water (20 ml) and ethyl acetate (30 ml), the crystalline product collected on a filter and washed with ethyl acetate (20 ml). The first crop thus obtained weighs 3.85 g. A second crop (0.56 g) is secured by washing the combined filtrates and washes with 0.5 N $KHCO_3$ (20 ml), water (20 ml), 0.5 N HCl (20 ml), water (20 ml) drying the solution over anhydrous Na_2SO_4, and evaporation in vacuo. The two crops are combined and recrystallized from water-saturated ethyl acetate by dilution with hexane. The purified protected dipeptide ester (4.2 g, 81%) melts at 155–156 °C; $[\alpha]_D^{22}$ −5 ° (*c* 2, chloroform).

1. Leplawy MT, Jones DS, Kenner GW, Sheppard RC (1960) Tetrahedron, 11: 39
2. Zaoral, M (1959) Angew Chem 71: 743; Collect Czechoslov Chem Commun 27: 1273 (1962)
3. Hofmann K, Thompson TA, Yajima H, Schwartz ET, Inouye H (1960) J Am Chem Soc 82: 3715
4. Prepared from glycine ethyl ester hydrochloride. A solution of the latter in water is covered with ether and cooled with ice water during the addition of a 50% solution of K_2CO_3. The ether extract is dried and the ether is evaporated under moderately reduced pressure (to avoid too heavy losses of the volatile ethyl glycinate).

4.3 The Ethyl Carbonate Mixed Anhydride Method [1]

Benzyloxy-carbonyl-L-leucyl-glycine Ethyl Ester [2]

$$C_6H_5-CH_2O-CO-NH-CH(CH_2-CH(CH_3)_2)-COO^-\cdot H\overset{+}{N}(C_2H_5)_3 + C_2H_5O-CO-Cl \longrightarrow$$

$$C_6H_5-CH_2O-CO-NH-CH(CH_2-CH(CH_3)_2)-CO-O-CO-OC_2H_5 + (C_2H_5)_3\overset{+}{N}H\cdot Cl^- \xrightarrow[N(C_2H_5)_3]{Cl^-\cdot H_3\overset{+}{N}CH_2CO-OC_2H_5}$$

$$C_6H_5-CH_2O-CO-NH-CH(CH_2-CH(CH_3)_2)-CO-NH-CH_2-CO-OC_2H_5 + (C_2H_5)_3\overset{+}{N}H\cdot Cl^- + C_2H_5OH + CO_2$$

$C_{18}H_{26}N_2O_5$ (350.4)

A one molar solution [3] of benzyloxycarbonyl-L-leucine in toluene (500 ml) is placed in a 3 liter round bottom flask protected from moisture by a cotton containing drying tube and diluted with chloroform (500 ml). The solution is stirred and neutralized with triethylamine [4] (50.6 g = 70 ml, 500 mmol). The mixture is cooled to about −15 °C and ethyl chlorocarbonate (about 97% pure, 50 ml = 56 g, 505 mmol) is added in 3 portions. Within a few minutes a mass of crystals separates. About 20 minutes later a precooled (ca −10 °C) solution of glycine ethyl ester hydrochloride (73 g, 525 mmol) and triethylamine (73 ml = 53 g, 525 mmol) in chloroform (1 liter) is added and stirring is continued at a bath temperature of −10 to 0 °C. When all the solid material dissolves the mixture is heated to 50 °C (CO_2 evolution) for about half an hour and then cooled to room temperature. The solution is washed with N HCl, H_2O, 0.5 M $KHCO_3$, again with H_2O, dried over $MgSO_4$, filtered and evaporated in vacuo to about 0.7 liter. On dilution with hexane (3 liters) the product separates in crystalline form. It is collected on a filter, washed with hexane (0.7 liter) and dried in air [5]. The protected dipeptide ester, 149 g (85%), m.p. 103–104 °C, $[\alpha]_D^{24}$ −26.8° (*c* 5, ethanol) is sufficiently pure for most practical purposes.

1. Boissonnas RA (1951) Helv Chim Acta 34: 874; Vaughan JR, Jr (1951) J Am Chem Soc 73: 3547; Wieland T, Bernhard H (1951) J Liebigs Ann Chem 572: 190; Vaughan JR, Jr, Osato RL (1952) J Am Chem Soc 74: 676
2. Bodanszky M, unpublished
3. A one ml aliquot of a solution of benzyloxycarbonyl-L-leucine in toluene is diluted with a few ml of ethanol and titrated with 0.1 N NaOH in the presence of phenolphthaleine.

4. Triethylamine is satisfactory in the activation and coupling of most amino acids protected by urethane type amine blocking groups. When conservation of chiral purity appears as a problem, for instance in connection with the activation and coupling of peptides, triethylamine should be replaced by *N*-methylmorpholine (Anderson GW, Zimmerman JE, Callahan FM (1966) J Am Chem Soc 88: 1338).
5. Drying in air should be carried out in a hood and the danger of fire (hexane vapors) should not be forgotten.

4.4 Isobutylcarbonic Acid Mixed Anhydrides [1]

Benzyloxy-carbonyl-glycyl-L-phenylalanyl-glycine Ethyl Ester [2]

$$C_6H_5-CH_2O-CO-NH-CH_2-CO-NH-CH(CH_2C_6H_5)-COOH + Cl-C(=O)-O-CH_2CH(CH_3)_2 + O(CH_2CH_2)_2N-CH_3 \longrightarrow$$

$$C_6H_5-CH_2O-CO-NH-CH_2-CO-NH-CH(CH_2C_6H_5)-CO-O-CO-O-CH_2CH(CH_3)_2 + O(CH_2CH_2)_2N-CH_3 \cdot HCl \xrightarrow[(C_2H_5)_3N]{H_2N-CH_2-CO-OC_2H_5 \cdot HCl}$$

$$C_6H_5-CH_2O-CO-NH-CH_2-CO-NH-CH(CH_2C_6H_5)-CO-NH-CH_2-CO-OC_2H_5 + CO_2 + (CH_3)_2CHCH_2OH + (C_2H_5)_3N\cdot HCl$$

$C_{23}H_{27}N_3O_6$ (441.5)

A solution of glycine ethyl ester hydrochloride (1.40 g, 10 mmol) in dimethylformamide (20 ml) is prepared by gentle warming. The solution is cooled to room temperature and treated with triethylamine [3] (1.01 g = 1.40 ml, 10 mmol).

A solution of benzyloxycarbonyl-glycyl-L-phenylalanine (3.56 g, 10 mmol) in dry tetrahydrofurane (50 ml) is cooled to $-15\,°C$ and neutralized with *N*-methylmorpholine (1.01 g – 1.10 ml, 10 mmol). Isobutyl chlorocarbonate [4] (1.37 g = 1.32 ml, 10 mmol) is added, followed, about one minute later, by the solution of glycine ethyl ester described above. About 2 ml dimethylformamide is used for rinsing. The reaction mixture is allowed to warm up to room temperature. The hydrochlorides of *N*-methylmorpholine and triethylamine are removed by filtration and washed with tetrahydrofurane. The combined filtrate and washings are concentrated in vacuo to about 25 ml and diluted with water (50 ml) [5]. The precipitate is collected on a filter and washed with water, 0.5 N $KHCO_3$, again with water. The dry product weighs 4.0 g (91%) and melts at 118–120 °C [6].

1. Vaughan JR Jr, Osato RL (1951) J Amer Chem Soc 73: 3547; 74: 676 (1952)
2. Anderson GW, Zimmerman JE, Callahan FM (1967) J Amer Chem Soc 89: 5012
3. Triethylamine can be replaced with the equivalent amount of *N*-methylmorpholine (1.01 g = 1.12 ml)
4. Commercially available as isobutyl chloroformate. Slightly better yields were reported (cf. ref. 1) with *sec*-butyl chlorocarbonate and, in recent years, with isopropyl chlorocarbonate.
5. Alternatively, the solvent is removed in vacuo, the residue dissolved in a mixture of ethyl acetate (150 ml) and water (50 ml) and the organic phase is washed with 0.5 N $KHCO_3$ (50 ml), 0.5 N HCl (50 ml), water (50 ml), dried over anhydrous Na_2SO_4 and evaporated to dryness in vacuo.
6. Fractional crystallization from 200 ml abs. ethanol yields no racemate (cf. Anderson GW, Callahan FM (1958) J Amer Chem Soc 80: 2092) and the recovered material (3.92 g, 84%) melts at 120–121 °C.

4.5 Coupling with the Aid of *o*-Phenylene Phosphorochloridite [1]

Preparation of *N*-benzyloxycarbonyl-*S*-benzyl-L-cysteinyl-L-tyrosyl-L-isoleucyl-L-glutaminyl-L-asparaginyl-*S*-benzyl-L-cysteinyl-L-prolyl-L-leucyl-glycine Amide [2]

Z–Cys(Bzl)–Tyr–Ile–Gln–Asn–OH + H–Cys(Bzl)–Pro–Leu–Gly–NH_2 + (o-phenylene)O_2P–Cl

$\xrightarrow{N(C_2H_5)_3}$ Z–Cys(Bzl)–Tyr–Ile–Gln–Asn–Cys(Bzl)–Pro–Leu–Gly–NH_2 +

$C_{65}H_{86}N_{15}O_{14}S_2$ (1323.5)

(o-phenylene)O_2P–$\bar{O}\cdot H\overset{+}{N}(C_2H_5)_3$ + $(C_2H_5)_3\overset{+}{N}H\cdot Cl^-$

The protected pentapeptide derivative *N*-benzyloxycarbonyl-*S*-benzyl-L-cysteinyl-L-tyrosyl-L-isoleucyl-L-glutaminyl-L-asparagine [2, 3] (8.7 g, 10 mmol) and the partially blocked tetrapeptide amide *S*-benzyl-L-cysteinyl-L-propyl-L-leucyl-glycinamide [2] (5.5 g, 11.5 mmol) are dissolved in dimethylformamide (50 ml) and the solution is cooled in an ice-water bath. Triethylamine (3.75 g = 5.2 ml, 37 mmol) is added with stirring followed by *o*-phenylene phosphorochloridite (3.75 g, 21.5 mmol). The mixture is allowed to warm up to room temperature and stirred overnight. The reaction mixture is diluted with ice water under vigorous stirring. Acetic acid (about 1 ml) is added to adjust the reaction of the suspension to slight acidity (ca pH 6). The precipitate is collected on a filter, thoroughly washed with water and dried in air. The crude product (12.2 g, m.p. 223–224 °C dec.) is suspended in methanol (300 ml) and the suspension is stirred at room temperature for about an hour. The insoluble material is filtered, washed with methanol, dried in air and finally in vacuo over phosphorus pentoxide. The protected nonapeptide derivative (with the sequence of oxytocin) weighs 9.6 g (72%) and melts at 235–236 °C dec. [4].

1. Anderson GW, Blodinger J, Young RW, Welcher AD (1952) J Amer Chem Soc 74: 5304; Anderson GW, Young RW (1952) ibid. 74: 5307; Anderson GW, Blodinger J, Welcher AD (1952) ibid. 74: 5390
2. Bodanszky M, du Vigneaud V (1959) J Amer Chem Soc 81: 2504
3. Activation of the carboxyl group of an asparagine residue is usually complicated by the formation of cyanoalanine derivatives. Yet, in the present example the expected product was obtained in fair yield and thus the dehydration reaction does not seem to interfere with the process to an unacceptable degree.
4. On deprotection by reduction with sodium in liquid ammonia, followed by air-oxidation of the resulting disulfhydro intermediate to oxytocin 200 units of biologically active material were obtained from 1 mg of the protected nonapeptide. This compares favorably with the results of several syntheses of oxytocin. When, however, instead of condensation of segments the stepwise strategy was applied (Bodanszky M, du Vigneaud V (1959) J Amer Chem Soc 81: 5688) the same protected nonapeptide was obtained with a higher melting point (245–248° dec.) and it yielded more biological activity on reduction and cyclization.

5 Preparation of Active Esters

5.1 Cyanomethyl Esters [1]

N-Benzyloxy-carbonyl-S-benzyl-L-cysteine Cyanomethyl Ester [2]

$$C_6H_5-CH_2O-CO-NH-CH(CH_2-S-CH_2-C_6H_5)-COOH + N(C_2H_5)_3 + ClCH_2CN \longrightarrow$$

$$C_6H_5-CH_2O-CO-NH-CH(CH_2-S-CH_2-C_6H_5)-CO-OCH_2CN + HCl \cdot N(C_2H_5)_3$$

$C_{20}H_{20}N_2O_4S$ (384.5)

A mixture of triethylamine (15.2 g = 21.0 ml, 150 mmol) and chloroacetonitrile (15.1 g = 12.6 ml, 200 mmol) is cooled in an ice-water bath while N-benzyloxycarbonyl-S-benzyl-L-cysteine [3] (34.5 g, 100 mmol) is added, in small portions, with stirring. The addition of the protected amino acid requires about 15 minutes. Stirring and cooling are continued for about 30 min and the mixture is stored at room temperature overnight. The thick mass is diluted with ethyl acetate (200 ml), the insoluble material (triethylammonium chloride) is removed by filtration and washed with ethyl acetate (50 ml). The solution is extracted with 0.5 N HCl (100 ml), 0.5 N $KHCO_3$ (100 ml) [4] and water (100 ml), dried over anhydrous Na_2SO_4 and evaporated to dryness in vacuo. The residue is crystallized from ether-hexane. The activated ester (31.2 g, 81%) melts at 65–67 °C; $[\alpha]_D^{19}$ −23° (c 4, $CHCl_3$); −45° (c 4, AcOH). Recrystallization from ether-hexane can raise the melting point to 67–68 °C.

1. Schwyzer R, Iselin B, Feurer M (1955) Helv Chim Acta 38: 69
2. Iselin B, Feurer M, Schwyzer R (1955) Helv Chim Acta 38: 1508
3. Harington CR, Mead TH (1936) Biochem J 30: 1598; Goldschmidt S, Jutz C (1953) Chem Ber 86: 1116

4. Acidification of the bicarbonate extract regenerates some (about 10%) of the protected amino acid.

5.2 Preparation of *p*-Nitrophenyl Esters [1]

Benzyloxy-carbonyl-L-phenylalanine *p*-Nitrophenyl Ester [2]

$C_6H_5CH_2$–C_6H_5–CH_2O–CO–NH–CH($CH_2C_6H_5$)–COOH + HO–C_6H_4–NO_2 + C_6H_{11}–N=C=N–C_6H_{11} ⟶

C_6H_5–CH_2O–CO–NH–CH($CH_2C_6H_5$)–C(=O)–O–C_6H_4–NO_2 + C_6H_{11}–NH–CO–NH–C_6H_{11}

$C_{23}H_{20}N_2O_6$ (420.4)

A solution of benzyloxycarbonyl-L-phenylalanine [3] (30.2 g, 101 mmol) and *p*-nitrophenol (16.7 g, 120 mmol) in ethyl acetate [4] (250 ml) is stirred with a magnetic stirrer and cooled in an ice-water bath. Dicyclohexylcarbodiimide [5] (20.6 g, 100 mmol) is added, in a few portions, through a powder funnel which is then rinsed with ethyl acetate (50 ml). After about 30 min the ice-water in the bath is replaced by water of room temperature and stirring continued for two more hours [6]. The *N*,*N'*-dicyclohexylurea which gradually separated is removed by filtration and thoroughly washed with ethyl acetate (500 ml) used in several portions [7]. The filtrate and washings are combined, evaporated in vacuo and the residue recrystallized from hot 95% ethanol containing 1% acetic acid [8]. The active ester, 31.5 g (75%) melts at 126–127 °C, $[\alpha]_D^{20}$ −24.7° (*c* 2, dimethylformamide [9, 10]).

1. For a general method of the synthesis of *p*-nitrophenyl esters cf. Bodanszky M, du Vigneaud V, (1963) Biochem Prep 9: 110
2. Bodanszky M, du Vigneaud V (1959) J Am Chem Soc 81: 6072
3. Benzyloxycarbonyl-L-phenylanine melts at 88–89 °C. A melting point of 126 to 128 °C is that of a complex which forms between the blocked amino acid and its sodium salt. This complex is not suitable for the preparation of active esters. It should be dissolved in water with the help of sodium carbonate and the solution poured, in a thin stream, into well stirred dilute ice-cold hydrochloric acid used in some excess. The precipitate is collected, washed with water and dried in air.
4. Ethyl acetate is a convenient solvent for the preparation of *p*-nitrophenyl esters, because it is volatile and not too toxic. Yet, other solvents such as pyridine, tetrahydrofurane or dimethylformamide are also suitable for the same purpose.
5. Dicyclohexylcarbodiimide is allergenic and should be handled with proper care. Traces of this material which adhere to glassware, spatulas, etc. should be decomposed, e.g. with dilute hydrochloric acid in aqueous ethanol, before cleaning.
6. Completeness of the esterification reaction could be ascertained by placing a few drops of the solution on the surface of a sodium chloride disc and after evaporation of the solvent recording the i.r. spectrum. The characteristic band of carbodiimides (2120 cm^{-1}) should be absent.

7. The product, *p*-nitrophenyl benzyloxycarbonyl-L-phenylalaninate, is only moderately soluble in ethyl acetate. Therefore if the air-dried byproduct (*N*,*N'*-dicyclohexylurea) weighs more than expected (ca 20 g), it contains some active ester which can be extracted with more ethyl acetate.
8. Acetic acid is added to prevent transesterification during recrystallization. Basic impurities, such as the sodium salt of the protected amino acid, catalyze transformation of the active ester to the ethyl ester.
9. Dimethylformamide usually contains enough basic impurities (probably mainly dimethylamine) to cause some decomposition of the active ester and the formation of nitrophenolate which results in a yellow color. Therefore, it is advisable to add about 1% acetic acid to the solvent before using it in the determination of specific rotation.
10. The same procedure has been applied to the preparation of other active esters as well; for instance 1-hydroxybenzotriazole esters of tritylamino acids (Barlos K, Papaioannov D, Theoropoulos D (1984) Int J Peptide Protein Res 23: 300, or esters of 3-hydroxy-3,4-dihydrobenzotriazinone-4-one of *N*-protected amino acids (König W, Geiger R (1970) Chem Ber 103: 2034; Atherton E, Holder JL, Meldal M, Sheppard RC, Valerio RM: J Chem Soc Perkin I. 1988: 2887).

5.3 *o*-Nitrophenyl Esters [1–3]

tert-Butyloxy-carbonyl-glycine *o*-Nitrophenyl Ester [4]

$(CH_3)_3C{-}O{-}CO{-}NH{-}CH_2{-}COOH$ + HO–C6H4–NO_2 (*o*) + C6H11–N=C=N–C6H11 ⟶

$(CH_3)_3C{-}O{-}CO{-}NH{-}CH_2{-}C(=O){-}O$–C6H4–$NO_2$ (*o*) + C6H11–NH–CO–NH–C6H11

$C_{13}H_{16}N_2O_6$ (296.3)

A solution of *tert*-butyloxycarbonyl-glycine (19.3 g, 110 mmol) and *o*-nitrophenol (27.8 g, 200 mmol) in pyridine [5] (300 ml) is cooled in an ice-water bath and stirred. Dicyclohexylcarbodiimide (20.6 g, 100 mmol) is added through a powder funnel and the latter is rinsed with pyridine (100 ml). Half an hour later the ice-water in the bath is replaced by water of room temperature and stirring is continued until no more carbodiimide can be detected by i.r. (2120 cm^{-1}) in a small sample of the solution. About four hours are needed to reach this point. The precipitate (*N*,*N'*-dicyclohexylurea) is filtered off and washed with pyridine. The solvent is removed in vacuo at a bath temperature not exceeding 25 °C. The residue is dissolved in ether (about 200 ml), the solution is filtered from some additional urea derivative and evaporated in vacuo. The residue is dissolved in chloroform (about 400 ml) and washed with a 5% solution of citric acid in water (200 ml in two portions), then several times with 0.1 N NaOH [6] and twice with water (200 ml each time). The organic layer is dried over $MgSO_4$ and evaporated in vacuo to dryness. The solid residue is recrystallized from hot 95% ethanol (300 ml). On cooling, the active ester separates in long white needles which are, at room temperature, somewhat soluble in ethanol. Therefore, crystallization is completed in the refrigerator, the crystals are collected on a filter [7] and washed with 95% ethanol cooled in ice-water. After drying in air the product weighs 22.9 g (77%) and melts at 96–98 °C. The active ester shows a characteristic [3] carbonyl band in the i.r. at 1780 cm^{-1}. On thin layer plates of silica gel it appears as a single spot with an R_f value of 0.63 in chloroform-methanol (9:1) [8]. A sample dried at room temperature in vacuo over phosphorus pentoxide gives the expected C, H and N values on elemental analysis.

1. Bodanszky M, Funk KW, Fink ML (1973) J Org Chem 38: 3565
2. Bodanszky M, Kondo M, Lin CY, Sigler GF (1974) J Org Chem 39: 444
3. Bodanszky M, Fink ML, Funk KW, Kondo M, Lin CY, Bodanszky A (1974) J Amer Chem Soc 96: 2234
4. Bodanszky M, Funk KW (1973) J Org Chem 38: 1296

5. The formation of *o*-nitrophenyl esters requires longer time than that of the corresponding *p*-nitrophenyl esters. This is probably due to a reduction in nucleophilic character of the hydroxyl group in *o*-nitrophenol because of intramolecular hydrogen bonding. Consequently the esterification procedure generally applied in the preparation of *p*-nitrophenyl esters (Bodanszky M, du Vigneaud V (1962) Biochem Prep 9: 110) gives less than satisfactory results, particularly in the synthesis of *o*-nitrophenyl esters of hindered amino acids, such as valine or isoleucine where *N*-acylurea derivatives can become major products. This problem is greatly alleviated by the application of pyridine instead of ethyl acetate as the solvent and by the use of *o*-nitrophenol in considerable excess. The protected amino acid is added, in the above procedure, in some excess in order to utilize the entire amount of the carbodiimide.
6. The first extracts are yellow: mainly the excess of the protected amino acid is removed. The following ones are red from sodium *o*-nitrophenolate. When the color changes to orange, extraction with alkali should be discontinued, since on prolonged contact with the organic layer it will hydrolyze a part of the active ester.
7. Preferably in a cold-room.
8. Some *o*-nitrophenyl esters, e.g. that of *tert*-butyloxycarbonyl-L-leucine, have low melting points and are too soluble in ethanol for recrystallization. These can be purified by chromatography on a silica gel column with chloroform as the sole eluent (cf. refs. 1 and 2).

5.4 2,4-Dinitrophenyl Esters [1]

N-Benzyloxycarbonyl-L-threonine 2,4-Dinitrophenyl Ester [2, 3]

C_6H_5–CH_2O–CO–NH–CH($CH(OH)CH_3$)–COOH + HO–$C_6H_3(NO_2)_2$ + C_6H_{11}–N=C=N–C_6H_{11} ⟶

C_6H_5–CH_2O–CO–NH–CH($CH(OH)CH_3$)–CO–O–$C_6H_3(NO_2)_2$ + C_6H_{11}–NH–CO–NH–C_6H_{11}

$C_{18}H_{17}N_3O_9$ (419.3)

A solution of benzyloxycarbonyl-L-threonine (2.53 g, 10 mmol) and 2,4-dinitrophenol (1.84 g, 10 mmol) in dry ethyl acetate (30 ml) is cooled in an ice-water bath while dicyclohexylcarbodiimide (2.06 g, 10 mmol) is added with stirring. After about 30 min at 0 °C the mixture is allowed to warm up to room temperature. Next day the precipitated dicyclohexylurea is removed by filtration and washed with dry ethyl acetate. The solvent is removed in vacuo and the residue crystallized from absolute ethanol by the addition of hexane. The crystals are washed on the filter with a 1:1 mixture of ethyl-acetate-hexane, then with hexane and are dried in vacuo. The active ester (3.58 g, 85%) melts at 90–92 °C; $[\alpha]_D^{20}$ −20° (*c* 2, dimethylformamide).

1. Bodanszky M (1955) Nature 175: 685. 2,4-Dinitrophenyl esters were obtained also via chlorides and mixed anhydrides of protected amino acids (Wieland T, Jaenicke F (1956) Justus Liebigs Ann Chem 599: 125; cf. also ref. 1). These highly reactive esters were prepared through the reaction of salts of carboxylic acids with di-2,4-dinitrophenyl carbonate as well (Glatthard R, Matter M (1963) Helv Chim Acta 46: 795).
2. Rocchi R, Marchiori F, Scoffone E (1963) Gazz Chim Ital 93: 823
3. The 2,4-dinitrophenyl esters of benzyloxycarbonyl-L-serine and benzyloxycarbonylnitro-L-arginine are similarly prepared (Bodanszky M, Ondetti MA, Chem Ind 1966: 26).

5.5 2,4,5-Trichlorophenyl Esters [1]

Tri-2,4,5-trichlorophenyl Phosphite

$$3\ \text{Cl}_3\text{C}_6\text{H}_2\text{–OH} + PCl_3 + N(C_2H_5)_3 \longrightarrow (\text{Cl}_3\text{C}_6\text{H}_2\text{–O–})_3P + 3\,Cl^- \cdot \overset{+}{N}H(C_2H_5)_3$$

$$\text{Z–NH–CHR–COOH} + (\text{Cl}_3\text{C}_6\text{H}_2\text{–O–})_3P \longrightarrow \text{Z–NH–CHR–CO–O–C}_6\text{H}_2\text{Cl}_3$$

2,4,5-Trichlorophenol [2] (59.3 g, 300 mmol) is dissolved in dry benzene [3] (500 ml), triethylamine (30.4 g = 42 ml, 300 mmol) is added and the solution is vigorously stirred during the *dropwise* [4] addition of phosphorus trichloride (13.8 g = 8.8 ml, 100 mmol). When the addition of PCl_3 is complete external heat is applied, the reaction mixture is boiled under reflux for 3 hours and then cooled to room temperature. The triethylammonium chloride is removed by filtration and the solvent evaporated in vacuo. The residue is recrystallized from hot ethyl acetate (130 ml) to afford tri-2,4,5-trichlorophenyl phosphite (43.4 g, 70%) melting at 120–121 °C. It has to be stored under the exclusion of moisture.

Esterification

The protected amino acid (10 mmol) is dissolved in dry pyridine (50 ml) and treated with tri-2,4,5-trichlorophenyl phosphite (50 g, 80 mmol). The mixture is shaken or vigorously stirred for 6 hours when the clear solution is evaporated to dryness in vacuo. The residue is dissolved in ethyl acetate, the solution extracted with 0.1 N HCl (50 ml), 0.5 N $KHCO_3$, a 10% solution of NaCl in water (50 ml), dried over anhydrous Na_2SO_4 and evaporated to dryness in vacuo. The crude product is recrystallized from ethanol.

1. Pless J, Boissonnas RA (1963) Helv Chim Acta 46: 1609
2. This compound is the precursor of the highly toxic dioxin. It must be handled with care.
3. Benzene is harmful. The operations described above should be carried out in a well ventilated hood.
4. There is a sharp rise in the temperature of the reaction mixture.

5.6 Pentachlorophenyl Esters [1]

Benzyloxy-carbonyl-glycyl-L-phenylalanine Pentachloro-phenyl Ester [2, 3]

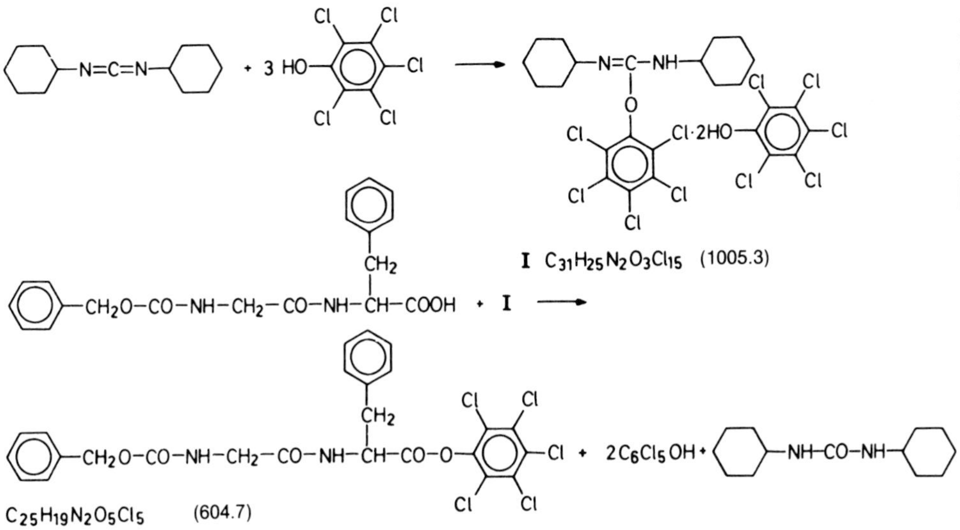

I $C_{31}H_{25}N_2O_3Cl_{15}$ (1005.3)

$C_{25}H_{19}N_2O_5Cl_5$ (604.7)

A solution of pentachlorophenol (8.8 g, 33 mmol) in ethyl acetate (10 ml) is cooled in an ice-water bath and dicyclohexylcarbodiimide (2.06 g, 10 mmol) is added with vigorous stirring. The mixture is stored at about −10 °C overnight. Some crystals deposit during that time. Ice cold hexane (10 ml) is added, the complex collected on a filter and washed with ice cold hexane. The crude product (I) is recrystallized from hot hexane (130 ml) to yield analytically pure *O*-pentachlorophenyl *N,N'*-dicyclohexylisourea pentachlorophenol complex (8.25 g, 82%) melting within a broad range (115–162 °C).

To a solution of the complex (I) (1.0 g, 1.0 mmol) in ethyl acetate (10 ml) benzyloxycarbonyl-glycyl-L-phenylalanine (3.56 g, 1.0 mmol) is added and the mixture is allowed to stand at room temperature overnight. During this time a thick mass of crystals forms. Dry ether (10 ml) is added to dilute the suspension and the mixture is cooled in ice-water for about one hour. To separate the ester from the accompanying dicyclohexylurea the filter-cake is suspended in dioxane (10 ml) refiltered and washed with dioxane (twice, 2 ml each time). The combined filtrate and washings are evaporated to dryness in vacuo, and the residue extracted once more with dioxane as before. The solvent is removed in vacuo, the crude material suspended in dry ether (10 ml) and kept at about −10 °C overnight. The active ester is filtered, washed with ice-cold dry ether and dried in vacuo. It weighs 0.47 g (78%) and melts at 159–162 °C [4]; $[\alpha]_D$ −37.9° (*c* 1, chloroform).

1. Kupryszewski G (1961) Roczniki Chem 35: 595; Chem Abstr 55: 27121i
2. Kovacs J, Kisfaludy L, Ceprini MQ, Johnson RH (1969) Tetrahedron 25: 2555
3. Pentachlorophenyl esters were prepared, in good yield, also by the reaction of protected amino acids with pentachlorophenyl trichloroacetate in pyridine (Fujino M, Hatanaka C (1968) Chem Pharm Bull 16: 929).
4. Higher melting points indicate chirally impure material (cf. ref. 2).

5.7 Pentafluorophenyl Esters [1]

Benzyloxy-carbonyl-glycyl-L-phenylalanine Pentafluoro-phenyl Ester [2]

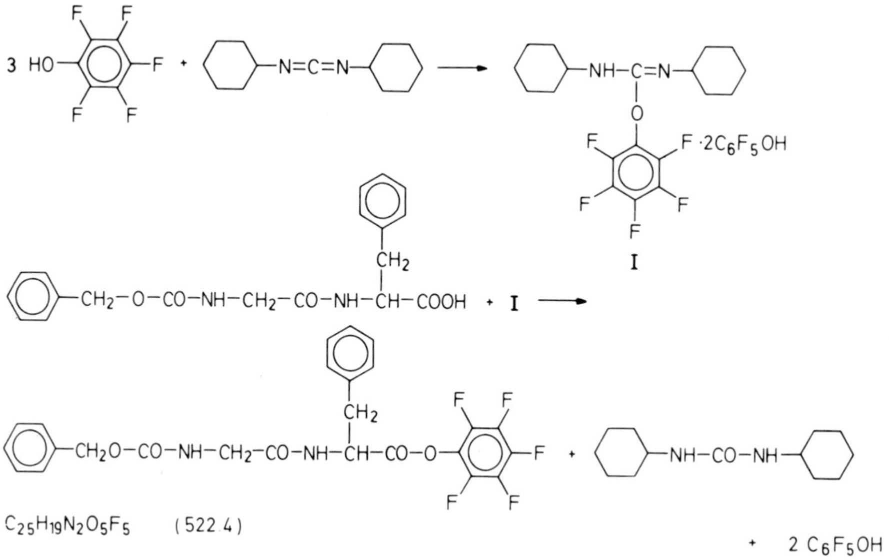

$C_{25}H_{19}N_2O_5F_5$ (522.4)

Pentafluorophenol [3] (5.52 g, 30 mmol) is added to an ice-cold solution of dicyclohexylcarbodiimide (2.06 g, 10 mmol) in ethyl acetate (40 ml) followed, in about 5 minutes, by the addition of an ice-cold solution of benzyloxycarbonyl-glycyl-L-phenylalanine (3.56 g, 10 mmol). The mixture is stirred at 0 °C for about one hour. The *N,N'*-dicyclohexylurea which separated is removed by filtration and the solution evaporated to dryness in vacuo. Ethyl acetate (20 ml) is added to the residue and the small amount of undissolved dicyclohexylurea is filtered off. Removal of the solvent in vacuo leaves the practically pure active ester (4.82 g, 92%) melting at 96–98 °C; $[\alpha]_D^{23}$ −10.5 (*c* 1, chloroform). Recrystallization from ethanol-water leaves the m.p. unchanged.

tert-Butyloxycarbonylamino Acid Pentafluorophenyl Esters [4]

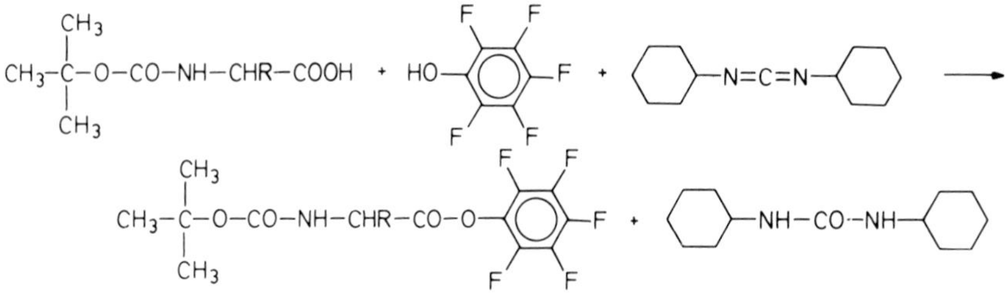

$$(CH_3)_3C{-}O{-}CO{-}NH{-}CHR{-}COOH + HO{-}C_6F_5 + C_6H_{11}{-}N{=}C{=}N{-}C_6H_{11} \longrightarrow (CH_3)_3C{-}O{-}CO{-}NH{-}CHR{-}CO{-}O{-}C_6F_5 + C_6H_{11}{-}NH{-}CO{-}NH{-}C_6H_{11}$$

A solution of the *tert*-butyloxycarbonylamino acid (10 mmol) and pentafluorophenol [3] (2.0 g, 11 mmol) in ethyl acetate [5] (10 ml) is cooled in an ice-water bath while dicyclohexylcarbodiimide (2.27 g, 11 mmol) is added with stirring. The stirred mixture is cooled in the same bath for one hour. The precipitated *N*,*N'*-dicyclohexylurea is filtered off and the solvent removed in vacuo. The residue solidifies on trituration with hexane. Pentafluorophenyl esters of *tert*-butyloxycarbonylamino acids can be stored in a desiccator for several weeks.

1. Kovacs J, Kisfaludy L, Ceprini MQ (1967) J Amer Chem Soc 89: 183
2. Kisfaludy L, Roberts JE, Johnson RH, Mayers GL, Kovacs J (1970) J Org Chem 35: 3563
3. Contact with the skin and particularly with the eyes should be avoided.
4. Kisfaludy L, Löw M, Nyéki O, Szirtes T, Schön I: Justus Liebigs Ann Chem 1973: 1421
5. For protected amino acids which are not readily soluble in ethyl acetate, dioxan or tetrahydrofuran can be used. The addition of a small amount of dimethylformamide greatly increases solubility.

5.8 *N*-Hydroxyphthalimide Esters [1, 2]

N-Hydroxyphthalimide (precursor)

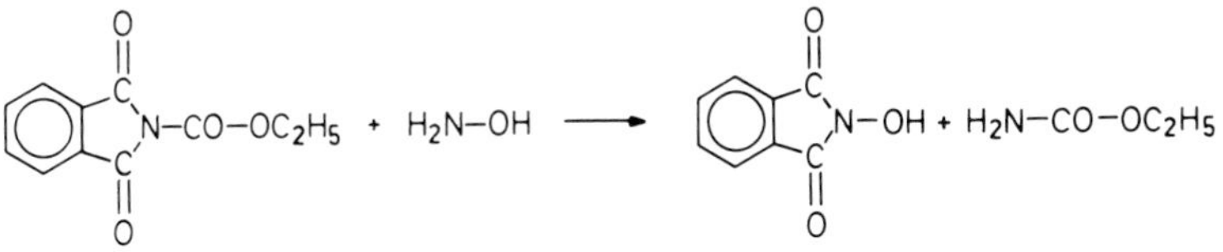

$$C_6H_4(CO)_2N{-}CO{-}OC_2H_5 + H_2N{-}OH \longrightarrow C_6H_4(CO)_2N{-}OH + H_2N{-}CO{-}OC_2H_5$$

$C_8H_5NO_3$ (163.1)

Esterification

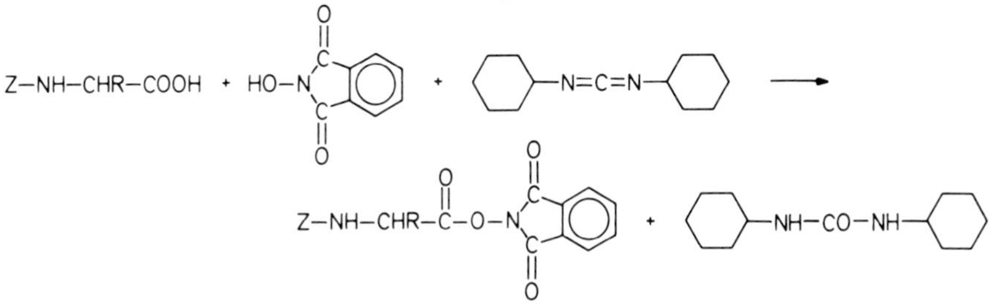

$$Z{-}NH{-}CHR{-}COOH + HO{-}N(CO)_2C_6H_4 + C_6H_{11}{-}N{=}C{=}N{-}C_6H_{11} \longrightarrow Z{-}NH{-}CHR{-}C(=O){-}O{-}N(CO)_2C_6H_4 + C_6H_{11}{-}NH{-}CO{-}NH{-}C_6H_{11}$$

A mixture of hydroxylamine hydrochloride (6.95 g, 100 mmol) and triethylamine (20.2 g = 28 ml, 200 mmol) in absolute ethanol (50 ml) is heated on a steam bath until complete solution occurs. *N*-Ethoxycarbonyl-phthalimide [3] (22 g, 100 mmol) is added in one portion to the hot solution with vigorous stirring. The solution takes up a deep red color. It is immediately cooled to room temperature and poured into a mixture of water (250 ml) and 2 N HCl (50 ml); *N*-hydroxyphthalimide separates in fine, colorless needles. The product is collected on a filter, washed with water and dried over P_2O_5 in vacuo. It weighs 11.4 g (70%) and melts at 230 °C.

The protected amino acid (10 mmol) is dissolved in dry dimethylformamide (about 8 ml) and *N*-hydroxyphthalimide (2.0 g, 12.3 mmol) is added. The solution is cooled to −5 °C and treated with a solution of dicyclohexylcarbodiimide (2.06 g, 10 mmol) in ethyl acetate (20 ml). The reaction mixture is stored at about 0 °C overnight. The separated *N,N'*-dicyclohexylurea is removed by filtration and washed with ethyl acetate. The combined filtrate and washings are extracted with 0.5 N $KHCO_3$ (twice, 10 ml each time), washed with water (10 ml), dried over anhydrous Na_2SO_4 and evaporated to dryness in vacuo. The residue is crystallized from ethanol or other appropriate solvent [4].

1. Nefkens, GHL, Tesser GI (1961) J Amer Chem Soc 83: 1263
2. Nefkens GHL, Tesser GI, Nivard RJF (1962) Rec Trav Chim Pay Bas 81: 683
3. Commercially available. Cf. Nefkens GHL (1960) Nature 185: 309; Nefkens GHL, Tesser GI, Nivards RJF (1960) Rec Trav Chim Pay Bas 79: 688
4. Ethyl acetate-hexane or ethyl acetate-toluene might be suitable for some *N*-hydroxyphthalimide esters.

5.9 *N*-Hydroxysuccinimide Esters [1]

tert-Butyloxycarbonyl-L-alanine N-Hydroxysuccinimide Ester [2]

$C_{12}H_{18}N_2O_6$ (286.3)

A solution of *tert*-butyloxycarbonyl-L-alanine (1.89 g, 10 mmol) and *N*-hydroxysuccinimide [3] (1.15 g, 10 mmol) in dry 1,2-dimethoxymethane [4] is cooled in an ice-water bath and dicyclohexylcarbodiimide (2.06 g, 10 mmol) is added with stirring. The mixture is kept in the refrigerator (between 0° and +5°) overnight. The separated *N,N'*-dicyclohexylurea is removed by filtration and the solvent evaporated in vacuo. The crude product is twice recrystallized from isopropanol. The analytically pure active ester (2.03 g, 71%) melts at 143–144 °C [5]; $[\alpha]_D^{25}$ −49° (*c* 2, dioxane) [6].

1. Anderson GW, Zimmerman JE, Callahan FM (1963) J Amer Chem Soc 85: 3039
2. Anderson GW, Zimmerman JE, Callahan FM (1964) J Amer Chem Soc 86: 1839
3. Commercially available. The pure material melts at 99–100° (cf. ref. 2). A m.p. below 95° indicates a significant amount of contamination, usually succinic acid mono-hydroxamic acid (HOOC–CH_2–CH_2–CO–NH–OH) formed by the action of moisture. The ring can be reclosed by storing the preparation over P_2O_5 in a vacuum desiccator at room temperature for a prolonged period or at 40–50 °C overnight; cf. Errera G (1895) Gazetta Chim Ital 25(II): 25.
4. Other solvents, e.g. dioxane could be equally suitable.
5. A second form melts at 167 °C.
6. Preparation of the hydroxysuccinimide ester of an *o*-nitrobenzenesulfenyl-amino acid is described on p. 207.

5.10 1-Hydroxypiperidine Esters [1]

Benzyloxy-carbonyl-L-leucine 1-Hydroxy-piperidine Ester [1]

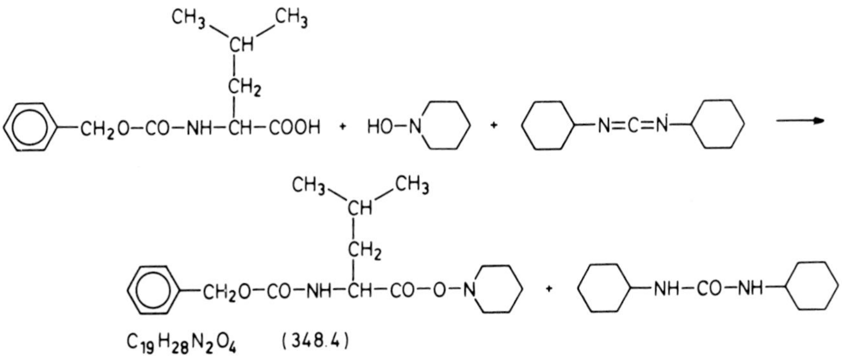

$C_{19}H_{28}N_2O_4$ (348.4)

A solution of benzyloxycarbonyl-L-leucine (2.65 g, 10 mmol) and 1-hydroxy-piperidine [2] (1.11 g, 11 mmol) in dry ethyl acetate [3] (50 ml) is stirred at room temperature and dicyclohexylcarbodiimide [4] (2.06 g, 10 mmol) is added. Six hours later the separated *N,N'*-dicyclohexylurea is removed by filtration and washed with ethyl acetate (20 ml). The combined filtrate and washings are extracted with N HCl [3] (twice, 30 ml each time), 0.5 N $KHCO_3$ (twice, 30 ml each time), with a saturated solution of NaCl in water (twice, 30 ml each time), dried over anhydrous Na_2SO_4 and evaporated to dryness in vacuo. Ether (20 ml) is added to the residue, a small amount of undissolved urea derivative is filtered off and the solution diluted with hexane until turbid.

A few drops of ether are added to restore the clarity of the solution which, on standing, deposits the crystalline ester. Crystallization is completed by storage in the refrigerator overnight. The crystals are collected on a filter, washed with an ice-cold mixture of ether-hexane (10 ml, mixed in a ratio in which the two solvents were used for crystallization) and finally with hexane. The product (2.90 g, 83%) melts at 65–67 °C. Recrystallization from ether-hexane raises the m.p. to 66–67 °C [3]; $[\alpha]_D^{24}$ $-11°$ (*c* 1, ethyl acetate); $[\alpha]_D^{20}$ $-20.6°$ (*c* 1, dimethylformamide). In the i.r. spectrum (mineral oil) the active ester carbonyl band appears at 1760 cm^{+1}.

1. Handford BO, Jones JH, Young GT, Johnson TFN: J Chem Soc 1965: 6814
2. Commercially available.
3. For the preparation of 1-piperidyl esters of *tert*-butyloxycarbonylamino acids ether should be used instead of ethyl acetate and N HCl applied in the work up should be replaced by a 10% solution of citric acid in water (Jones JH, Young GT, J Chem Soc (C) 1968: 53).
4. This ester was prepared (in 60% yield) also via a mixed anhydride. It melted, after recrystallization from diisopropyl ether at 67–68 °C. This is, however, a dangerous solvent, which might explode.

5.11 Esters of 5-Chloro-8-Hydroxy-Quinoline [1]

Benzyloxy-carbonylamino acid-5-Chloro-8-Hydroxy quinoline Esters

$$\text{Z–NH–CHR–COOH} + \text{Cl–C(=O)–OC}_2\text{H}_5 + \text{N(C}_2\text{H}_5)_3 \longrightarrow \text{Z–NH–CHR–CO–O–CO–OC}_2\text{H}_5$$

$$\xrightarrow{\text{5-chloro-8-hydroxyquinoline}} \text{Z–NH–CHR–C(=O)–O–C}_9\text{H}_5\text{N–Cl} + CO_2 + C_2H_5OH$$

A solution of the benzyloxycarbonylamino acid (10 mmol) in dry tetrahydrofurane (50 ml) is cooled to -15 °C and ethyl chlorocarbonate [2, 3] (1.09 g = 0.97 ml, 10 mmol) is added dropwise with stirring. Ten minutes later a solution of 5-chloro-8-hydroxy-quinoline [4] (1.80 g, 10 mmol) in tetrahydrofurane (20 ml) is added and the reaction mixture is stirred at -15 °C for one hour and then at room temperature for an additional hour. Water (20 ml) is added and the tetrahydrofurane is removed in vacuo. After the addition of ethyl acetate (50 ml) the organic layer is extracted with 0.5 N $KHCO_3$ (30 ml), water (30 ml), 0.5 N HCl (30 ml), water (30 ml), dried over anhydrous Na_2SO_4 and evaporated to dryness in vacuo. The residue, the 5-chloro-8-hydroxyquinoline ester of the appropriate benzyloxycarbonylamino acid [5] is obtained in 79 to 84% yield.

1. Jakubke HD, Voigt A (1966) Chem Ber 99: 2944
2. Usually sold under the name ethyl chloroformate.
3. Activation with dicyclohexylcarbodiimide afforded active esters of 5-chloro-8-hydroxy-quinoline in somewhat lower yield.
4. Commercially available. It is an irritant.
5. Physical properties of several benzyloxycarbonylamino acid 5-chloro-8-hydroxyquinoline esters are listed in ref. 1.

6 Coupling with Active Esters

6.1 Coupling in Organic Solvents

6.1.1 Acylation with Cyanomethyl Esters [1]

N-Benzyloxy-carbonyl-S-benzyl-L-cysteinyl-L-tyrosine

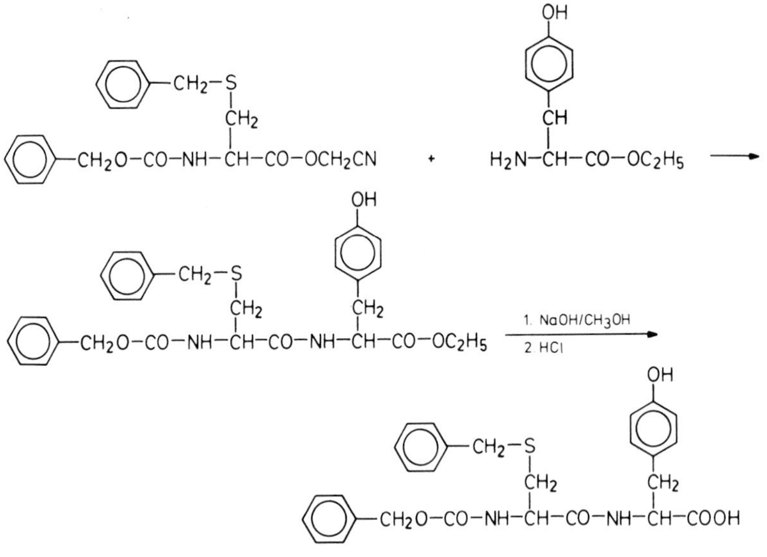

$C_{27}H_{28}N_2O_6S$ (508.6)

L-Tyrosine ethyl ester [2] (2.30 g, 11 mmol) is dissolved, with gentle warming, in dry tetrahydrofurane (5 ml). The solution is cooled to room temperature and N-benzyloxycarbonyl-S-benzyl-L-cysteine cyanomethyl ester [1, 3] (3.84 g, 10 mmol) is added followed by a catalytic amount (0.03 ml) of acetic acid. The reaction is allowed to proceed at room temperature for two days. The solvent is removed in vacuo, the residue dissolved in ethyl acetate (100 ml) and the solution extracted with 0.5 N $KHCO_3$ (50 ml), water (50 ml), 0.5 N HCl (50 ml) and water (50 ml), dried over anhydrous Na_2SO_4 and evaporated to dryness in vacuo.

The residue (6.1 g) is dissolved in a mixture of methanol (20 ml) and N NaOH (25 ml) and the solution kept at room temperature for 2 hours. After the addition of N HCl (10 ml) the methanol is removed in vacuo, the remaining aqueous solution cooled with ice-water and acidified with 6 N HCl to Congo. The precipitate is collected on a filter, washed with water and dried in vacuo. The crude material melts at 185–189 °C. Recrystallization from aqueous ethanol affords the title compound (3.6 g, 71%) in analytically pure form melting at 199–201 °C; $[\alpha]_D^{22} -17°$ (c 4, pyridine).

1. Iselin B, Feurer M, Schwyzer R (1955) Helv Chim Acta 38: 1508
2. Fischer E (1901) Ber Dtsch Chem Ges 34: 433; cf. p. 29
3. Cf. this volume p. 96

6.1.2 Chain Lengthening with *p*-Nitrophenyl Esters

Benzyloxy-carbonyl-L-asparaginyl-S-benzyl-L-cysteinyl-L-prolyl-L-leucyl-glycinamide [1]

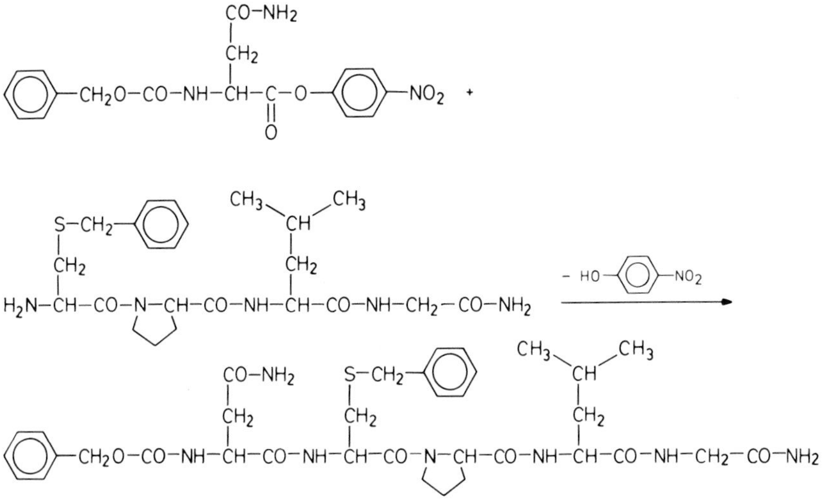

$C_{35}H_{47}N_7O_8S$ (725.8)

S-Benzyl-L-cysteinyl-L-prolyl-L-leucyl-glycinamide [2, 3] (hydrate, 5.0 g, 10 mmol) is suspended in ethyl acetate [4] (100 ml). The *p*-nitrophenyl ester of benzyloxycarbonyl-L-asparagine [5] (4.65 g, 12 mmol) is added in finely powdered form and the suspension is stirred at room temperature for two days. The starting materials gradually dissolve while the protected pentapeptide derivative separates. It is collected on a filter [6], washed with ethyl acetate (100 ml), ethanol (50 ml) and dried in air. The yield (7.25 g) is quantitative. The product melts at 212–213 °C and decomposes at 214 °C; $[\alpha]_D^{22} -59°$ (c 1, dimethylformamide) it gives satisfactory analytical values without further purification [7].

1. Bodanszky M, du Vigneaud V (1959) J Am Chem Soc 81: 5688
2. Ressler C, du Vigneaud V (1954) J Am Chem Soc 76: 3107

3. Zaoral M, Rudinger J (1955) Coll Czech Chem Commun 20: 1183
4. Similarly to several other active esters *p*-nitrophenyl esters react much faster in dimethylformamide than in ethyl acetate. The tendency, however, of benzyloxycarbonyl-L-asparagine *p*-nitrophenyl ester to produce benzyloxycarbonylaminosuccinimide is enhanced in the more polar solvent, while it is not pronounced in ethyl acetate. The unusually long reaction time, in this case, is due to the poor solubility of both reactants in this solvent.
5. Bodanszky M, Denning GS, Jr, du Vigneaud V (1963) Biochem Prep 10: 122
6. A sinter-glass filter should be used and the filter cake evenly packed with the help of a sturdy glass rod ending in a button-like flat head. It is necessary to pay attention to such small technical details because otherwise the intermediates cannot be secured directly in satisfactory purity and additional purification is needed. Yet, the poor solubility of many protected peptide intermediates in organic solvents precludes their efficient purification. Hence, it can be crucial to secure protected peptides as pure as possible, particularly in stepwise chain lengthening.
7. In this acylation reaction the amino compound is applied as the free base. Quite frequently it is more practical to use a mixture of a salt of the amino component with a tertiary amine such as triethylamine, *N*-methylpiperidine, *N*-methylmorpholine or diisopropylethylamine. Yet, better acylation rates can be expected in the acylation of free amines and there is less risk of racemization if no tertiary base is present in the reaction mixture.

6.1.3 Acylation with *N*-Hydroxysuccinimide Esters [1]

tert-Butyloxycarbonyl-glycyl-L-phenylalanyl-glycine Ethyl Ester

$$(CH_3)_3C{-}O{-}CO{-}NH{-}CH_2{-}CO{-}O{-}N(CO{-}CH_2{-}CH_2{-}CO) + H_2N{-}CH(CH_2C_6H_5){-}CO{-}NH{-}CH_2{-}CO{-}OC_2H_5 \cdot HBr + N(C_2H_5)_3 \longrightarrow$$

$$(CH_3)_3C{-}O{-}CO{-}NH{-}CH_2{-}CO{-}NH{-}CH(CH_2C_6H_5){-}CO{-}NH{-}CH_2{-}CO{-}OC_2H_5 + HBr \cdot N(C_2H_5)_3 + HO{-}N(CO{-}CH_2{-}CH_2{-}CO)$$

$C_{20}H_{29}N_3O_6$ (407.5)

L-Phenylalanyl-glycine ethyl ester hydrobromide (3.31 g, 10 mmol), *tert*-butyloxycarbonyl-glycine *N*-hydroxysuccinimide ester (2.72 g, 10 mmol) and triethylamine (1.01 g = 1.40 ml, 10 mmol) are added to dry 1,2-dimethoxyethane (33 ml), the mixture is stirred at room temperature for about 20 minutes, then poured into ice-cold water [2] (130 ml) with stirring. The material which separates is allowed to solidify. The crude product is collected on a filter, thoroughly washed with water and dried. It weighs 3.44 g (85%) and melts at 98–99 °C. Recrystallization from ethanol-water affords the pure tripeptide derivative (3.2 g, 79%) melting at 100.5–101.5 °C; $[\alpha]_D^{25} -10°$ (*c* 2, methanol).

Benzyloxycarbonyl-L-phenylalanyl-L-tyrosine Ethyl Ester

$C_{28}H_{30}N_2O_6$ (490.6)

A solution of benzyloxycarbonyl-L-phenylalanine *N*-hydoxysuccinimide ester (3.96 g, 10 mmol) in 1,2-dimethoxyethane (20 ml) is added to a stirred solution of L-tyrosine ethyl ester [3] (2.09 g, 10 mmol) in 1,2-dimethoxyethane (20 ml). Forty minutes later the mixture is diluted with water (120 ml), the precipitate collected on a filter and washed with a 10% solution of Na_2CO_3 in water (40 ml), water (40 ml), N HCl (40 ml) and water (40 ml). The dried material is recrystallized from ethanol water. After a second recrystallization it melts at 156–158 °C; $[\alpha]_D^{25} -9.1$ (*c* 10, ethanol) [4]. The yield is 4.17 g (85%).

1. Anderson GW, Zimmerman JE, Callahan FM (1964) J Amer Chem Soc 86: 1839
2. The by-products, triethylammonium bromide and *N*-hydroxysuccinimide, are soluble in water. In the similarly efficient acylation with *N*-hydroxyphthalimide esters (Nefkens GHL, Tesser GI, Nivard RJF (1962) Rec Trav Chim Pay-Bas 81: 683) an aqueous solution of sodium bicarbonate is used to dissolve the eliminated *N*-hydroxyphthalimide.
3. Fischer E (1901) Ber Dtsch Chem Ges 34: 4333; cf. p. 29.
4. The same compound was reported with the melting point of 159–160 °C and specific rotation $[\alpha]_D^{24} -9.1$ (*c* 10, ethanol) in the literature; cf. Vaughan JR, Jr., Osato R (1956) J Amer Chem Soc 74: 676

6.1.4 Coupling with 2,4-Dinitrophenyl Esters [1]

Benzyloxy-carbonyl-L-threonyl-L-alanyl-L-alanyl-L-alanine Ethyl Ester

$$C_6H_5-CH_2O-CO-NH-CH(CH(OH)CH_3)-C(=O)-O-C_6H_3(NO_2)_2 + H_2NCH(CH_3)CO-NHCH(CH_3)CO-NHCH(CH_3)CO-OC_2H_5$$

$$\longrightarrow C_6H_5-CH_2O-CO-NHCH(CH(OH)CH_3)CO-NHCH(CH_3)CO-NHCH(CH_3)CO-NHCH(CH_3)CO-OC_2H_5$$

$C_{23}H_{34}N_4O_8$ (494.5)

Triethylamine (2.43 g=3.36 ml, 24 mmol) is added to a well stirred suspension of L-alanyl-L-alanyl-L-alanine ethylester hydrobromide (4.08 g, 12 mmol) in dry chloroform (50 ml). When complete solution occurs benzyloxycarbonyl-L-threonine 2,4-dinitrophenyl ester [2] (4.19 g, 10 mmol) is added and the mixture is stirred at room temperature overnight. During this time a gelatineous precipitate forms. The solvent is removed in vacuo and ethyl acetate (50 ml) is added to the residue. The precipitate [3] is collected on a filter, washed with ethyl acetate (50 ml in several portions), then with a 5% solution of Na_2CO_3 in water (50 ml in several portions), water (20 ml), N HCl (50 ml in several portions) and finally with water until the washings are practically neutral. The product is dried over P_2O_5 in vacuo, dissolved in hot ethanol, the solution decolorized with activated charcoal, concentrated to a small volume and the tetrapeptide derivative precipitated by the addition of dry ether. The analytically pure product (3.7 g, 75%) melts at 201–202 °C; $[\alpha]_D^{20} -12.5°$ (*c* 2, dimethylformamide).

1. Rocchi R, Marchiori F, Scoffone E (1963) Gazzetta Chim Italiana 93: 823
2. Cf. pp. 99–100.
3. The precipitate consists of a mixture of the target tetrapeptide derivative and triethylammonium salts.

6.2 Coupling in Organic-Aqueous Media

6.2.1 Acylation with *p*-Nitrophenyl Esters

Benzyloxy-carbonyl-L-phenylalanyl-glycyl-L-proline [1]

$C_6H_5-CH_2O-CO-NH-CH(CH_2C_6H_5)-CO-O-C_6H_4-NO_2 + H_2N-CH_2-CO-N-CH-COONa$ (proline) $\xrightarrow[2\ HCl]{1\ NaOH}$

$C_6H_5-CH_2O-CO-NH-CH(CH_2C_6H_5)-CO-NH-CH_2-CO-N-CH-COOH$ (proline) $+ HO-C_6H_4-NO_2$

$C_{24}H_{27}N_3O_6$ (453.5)

A sample of glycyl-L-proline [2] (18.6 g, 108 mmol) is dissolved in distilled water (160 ml) and the solution diluted with pyridine (160 ml). The pH of the well stirred mixture is brought into the range of 8.5–9.0 by the addition of 5 N NaOH. Stirring is continued at room temperature while benzyloxycarbonyl-L-phenylalanine *p*-nitrophenyl ester [3] (42.0 g, 100 mmol) is added in several portions. The alkalinity of the mixture is maintained in the pH range 8.5–9.0 by the addition of 5 N NaOH [4, 5]. About an hour is required for the addition of the reactants. Stirring is continued for an additional hour. During the reaction the active ester gradually dissolves. Distilled water (250 ml) and sodium hydrogen carbonate (50 g) are added and the solution extracted with ethyl acetate (800 ml, in eight portions). The aqueous layer is acidified with 6 N HCl to Congo: an oil separates and soon solidifies. The crystalline mass is disintegrated with the help of a sturdy glass rod, collected on a filter and thoroughly washed with water. The air-dried product weighs 41 g (90%) and melts at 205–206 °C. Recrystallization from boiling ethanol raises the m.p. to 212 °C; $[\alpha]_D^{20}$ −62.6° (*c* 1.0, dimethylformamide) [6].

1. Bodanszky M, Sheehan JT, Ondetti MA, Lande S (1963) J Amer Chem Soc 85: 991
2. This dipeptide (m.p. 184 °C) is obtained by the hydrogenation of benzyloxy-carbonyl-glycyl-L-proline in aqueous ethanol, in the presence of a Pd on charcoal catalyst. The procedure described above for the preparation of the title tripeptide derivative is suitable also for the acylation of L-proline with benzyloxycarbonyl-glycine *p*-nitrophenyl ester.
3. Preparation of this active ester is described on page 97.
4. A total of about 36 ml of 5 N NaOH (180 mequ) is needed.
5. The reaction can be carried out on a pH-stat.
6. On elemental analysis the expected C, H and N values are obtained.

6.2.2 Acylation with *N*-Hydroxysuccinimide Esters [1]

Benzyloxy-carbonyl-glycyl-L-proline

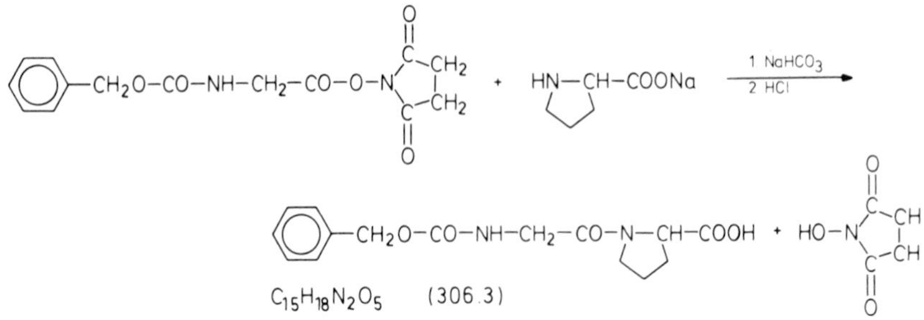

$C_{15}H_{18}N_2O_5$ (306.3)

A solution of L-proline (1.73 g, 15 mmol) and sodium hydrogen carbonate (1.26 g, 15 mmol) in water (16 ml) is treated with a solution of benzyloxy-carbonyl-glycine *N*-hydroxysuccinimide ester (3.06 g, 10 mmol) in 1,2-dimethoxyethane (20 ml). One hour later water (10 ml) is added and the solution is acidified to pH 2 with conc. hydrochloric acid. The mixture is cooled in an ice-water bath for about 30 min, the crystals which separated are collected and washed with a small volume of cold water. On further cooling a second crop forms. The two crops are combined (2.26 g) and recrystallized from hot ethyl acetate. The purified dipeptide derivative (1.94 g) melts at 158–159 °C. A second crop forms on prolonged cooling (0.36 g) and melts at 147–158 °C. The total yield is 75%.

Benzyloxy-carbonyl-L-prolyl-glycine

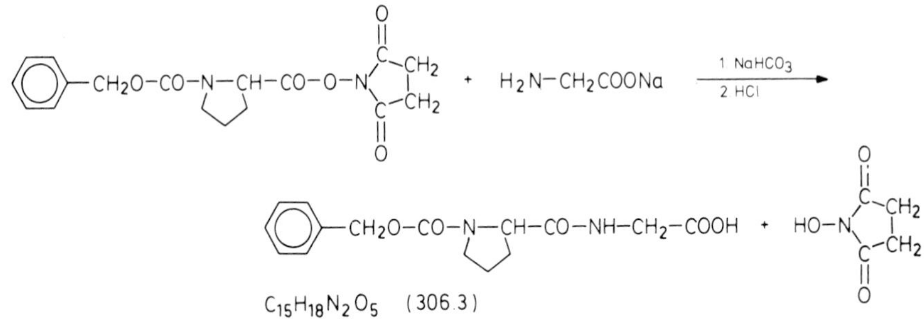

$C_{15}H_{18}N_2O_5$ (306.3)

Benzyloxycarbonyl-L-proline *N*-hydroxysuccinimide ester (3.46 g, 10 mmol) is dissolved in absolute ethanol [2] (34 ml) and the solution is added to a solution of glycine (0.75 g, 10 mmol) and $NaHCO_3$ (1.68 g, 20 mmol) in water (24 ml). The mixture is stirred at room temperature overnight, concentrated to a small volume in vacuo and acidified to pH 2 with conc. hydrochloric acid. The oil which separates changes into crystals and crystallization is completed by cooling the mixture in an ice-water bath. The crystals are washed on a filter with a small volume of cold water and dried. The crude dipeptide acid (2.8 g) melts at 116–124 °C. Recrystallization from ethyl acetate-hexane raises the melting point to 124–125 °C; the yield is 73% (2.24 g).

1. Anderson GW, Zimmerman JE, Callahan FM (1964) J Amer Chem Soc 86: 1839
2. The solution should be prepared without heating and should be used immediately.

6.3 Catalysis of Active Ester Reactions

6.3.1 Catalysis with Imidazole [1]

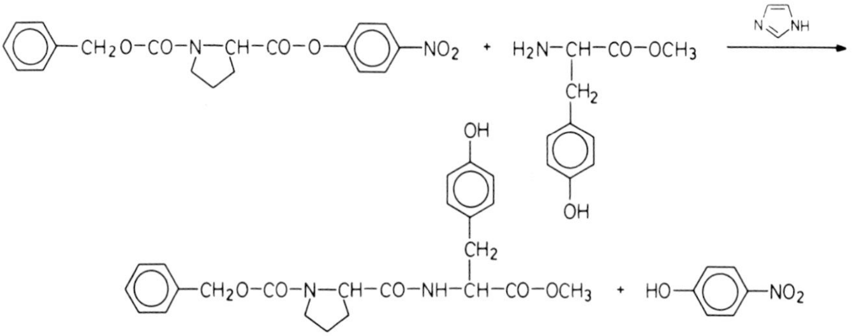

Benzyloxy-carbonyl-L-prolyl-L-tyrosine Methyl Ester

$C_{23}H_{26}N_2O_6$ (426.5)

To a solution of L-tyrosine methyl ester [2] (2.34 g, 12 mmol) and imidazole (6.8 g, 100 mmol) in ethyl acetate (125 ml) benzyloxycarbonyl-L-proline *p*-nitrophenyl ester [3] (3.70 g, 10 mmol) is added and the mixture is kept at room temperature for one hour. The solution is extracted with 0.5 N HCl (four times, 50 ml each time), 0.5 N NH_4OH (6 times, 50 ml each time) and water (50 ml), dried over anhydrous Na_2SO_4 and evaporated to dryness in vacuo. The residue is dissolved in ethyl acetate (15 ml) and cyclohexane (15 ml) is added. Crystals form which are collected on a filter, washed with a 1:1 mixture of ethyl acetate-cyclohexane (10 ml), with cyclohexane (10 ml) and dried in vacuo. The analytically pure benzyloxycarbonyl-L-prolyl-L-tyrosine methyl ester (3.88 g, 91%) melts at 82–85 °C; $[\alpha]_D^{26} -26°$ (*c* 1, methanol).

1. Mazur RH (1964) J Org Chem 28: 2498; cf. also Wieland T, Vogeler K (1964) Justus Liebigs Ann Chem 680: 125
2. Fischer, E (1908) Ber Dtsch Chem Ges 41: 850
3. Bodanszky M, du Vigneaud V (1959) J Amer Chem Soc 81: 5688

6.3.2 Catalysis with 1-Hydroxybenzotriazole [1]

Benzyloxy-carbonyl-L-leucyl-L-valinamide

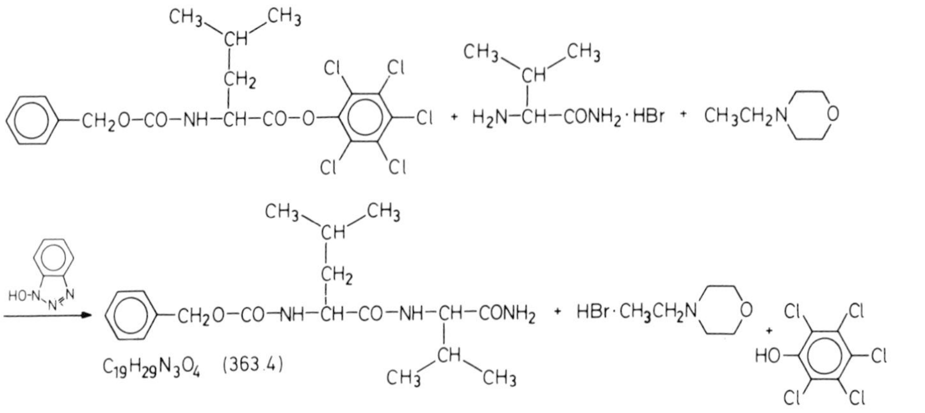

$C_{19}H_{29}N_3O_4$ (363.4)

To a solution of valinamide hydrobromide [2] (2.17 g, 11 mmol) and 1-hydroxybenzotriazole (monohydrate, 1.68 g, 11 mmol) in dimethylformamide (30 ml) *N*-ethylmorpholine (1.33 g = 1.45 ml, 11.5 mmol) is added followed by benzyloxycarbonyl-L-leucine pentachlorophenyl ester [3] (5.14 g, 10 mmol). The mixture is stirred at room temperature for about ten minutes then the solvent is removed in vacuo at a bath temperature not exceeding 30 °C. The residue is triturated with a saturated solution of $NaHCO_3$ in water (50 ml), the solid material collected on a filter, washed with water (50 ml) resuspended in a 2.5% solution of $KHSO_4$ in water (50 ml), filtered and thoroughly washed on the filter with water (100 ml). The product, dried in vacuo at 50 °C over P_2O_5 weighs 3.55 g (97.5%) and melts at 236–237 °C; $[\alpha]_D^{20} + 6.7°$ (*c* 1, dimethylacetamide). On elemental analysis correct values are found for C, H and N.

1. König W, Geiger R (1973) Chem Ber 106: 3626
2. Bodanszky M, Williams NJ (1967) J Amer Chem Soc 89: 685
3. Kovacs J, Ceprini MQ, Dupraz CA, Schmidt GN (1967) J Org Chem 32: 3696

6.3.3 Catalysis with 3-Hydroxy-3,4-dihydro-quinazoline-4-one [1]

Benzyloxy-carbonyl-L-isoleucyl-glycinamide

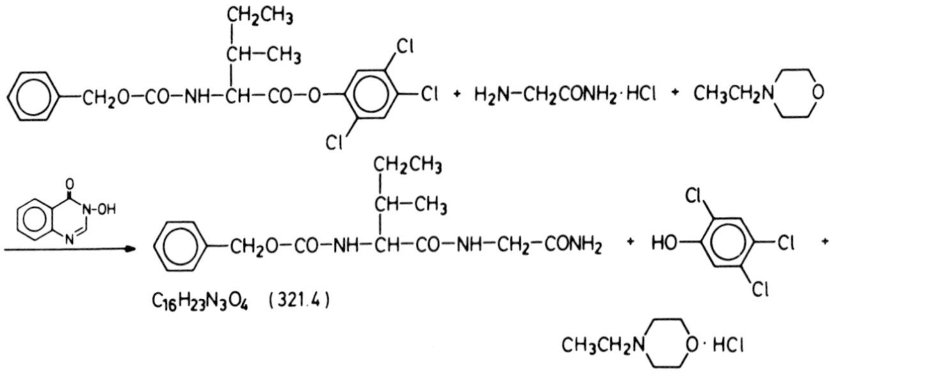

$C_{16}H_{23}N_3O_4$ (321.4)

N-Ethylmorpholine (11.5 g = 12.7 ml, 100 mmol) is added to a solution of benzyloxycarbonyl-L-isoleucine 2,4,5-trichlorophenyl ester [2] (4.45 g, 10 mmol), glycinamide hydrochloride (1.11 g, 10 mmol) and 3-hydroxy-3,4-dihydro-quinazoline-4-one [3] (0.16 g, 1 mmol) in dimethylformamide (75 ml). The solvent is removed in vacuo at a bath temperature of 30 °C and the residue triturated with 0.5 N $KHCO_3$ (30 ml). The solid material is collected on a filter, thoroughly washed with water and dried. The crude product is dissolved in dimethylformamide and precipitated with a 1:1 mixture of ether and hexane. The protected dipeptide amide (2.84 g, 88%) melts at 202 °C; $[\alpha]_D^{20}$ +8.1° (*c* 1, dimethylacetamide).

1. König W, Geiger R (1973) Chem Ber 106: 3626
2. Pless J, Boissonnas RA (1963) Helv Chim Acta 46: 1609
3. Harrison D, Smith ACB J Chem Soc 1960: 2157; cf. page 200

7 Peptide Bond Formation with the Aid of Coupling Reagents

7.1 The Dicyclohexylcarbodiimide (DCC, DCCI) Method [1]

7.1.1 Coupling with DCC

N-Phthaloyl-L-threonyl-L-phenylalanine Methyl Ester [2]

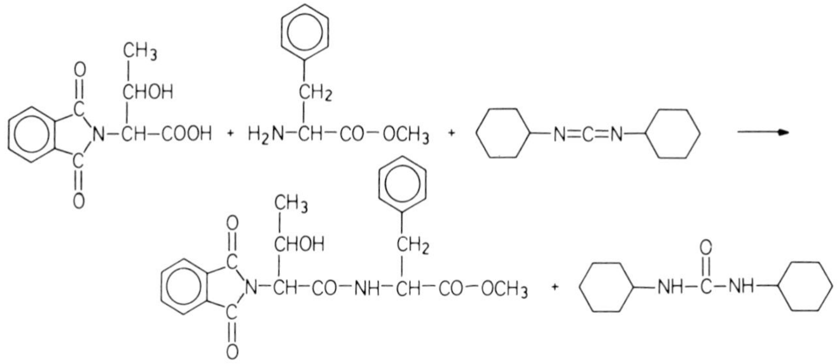

$C_{22}H_{22}N_2O_6$ (410.4)

A solution of L-phenylalanine methyl ester hydrochloride (5.4 g, 25 mmol) in water (20 ml) is treated with a solution of K_2CO_3 (5.0 g, 36 mmol) in water (10 ml) and the mixture is extracted with ether (three times, 25 ml each time). The ether extracts are pooled, dried over $MgSO_4$ and the solvent removed under moderately reduced pressure with the help of a water aspirator. The residue [3] (about 3.6 g, 20 mmol) is added to a solution of *N*-phthaloyl-L-threonine [4] (2.5 g, 10 mmol) in dichloromethane [5] (40 ml) followed by the addition of dicyclohexylcarbodiimide [6, 7] (2.06 g, 10 mmol). A precipitate, *N,N'*-dicyclohexylurea, starts to separate almost immediately and its amount gradually increases. After five hours at room temperature the urea derivative is removed by filtration [8] and washed with dichloromethane (20 ml). The combined filtrate and washings are extracted with N HCl (30 ml), N $KHCO_3$ (30 ml), water (30 ml), dried over $MgSO_4$ and evaporated to dryness in vacuo. Recrystallization of the residue from acetone-ether affords the pure dipeptide derivative (3.72 g, 91%) melting at 149–152 °C. A sample is recrystallized for analysis from acetone-ether melts at 153–154 °C; $[\alpha]_D^{27} + 1.9°$ (*c* 0.7, dimethylformamide).

1. Sheehan JC, Hess GP (1955) J Amer Chem Soc 77: 1067
2. Sheehan JC, Goodman M, Hess GP (1956) J Amer Chem Soc 78: 1367
3. The free amino acid ester should not be stored: it is gradually transformed into the diketopiperazine cyclo-L-Phe-L-Phe-.
4. Hydroxyamino acids can be used without blocking the hydroxyl group, but with excess acylating agent, *O*-acylation can occur; cf. Bodanszky M, Ondetti MA Chem Ind 1966: 26.
5. Preferably freshly distilled solvent should be used.
6. Equally good results can be obtained with diisopropylcarbodiimide.
7. Carbodiimides are known for causing allergic reactions. They should be handled with care; contact with the skin and particularly with the eyes should be avoided.
8. Removal of *N*, *N'*-dicyclohexylurea is usually incomplete; a small amount which remains in solution can contaminate the product and has to be removed from it by crystallization or chromatography. This difficulty can be circumvented by the application of water soluble carbodiimides since they give rise to water soluble urea derivatives. (cf. Sheehan JC, Hlavka JJ (1956) J Org Chem 21: 439). The water soluble reagent *N*-ethyl-*N'*-3-dimethylamino-propylcarbodiimide hydrochloride is commercially available.

7.1.2 Coupling with Carbodiimides in the Presence of 1-Hydroxybenzotriazole [1]

tert-Butyloxy-carbonyl-L-leucyl-L-phenylalanine Methyl Ester

$$(CH_3)_3C{-}O{-}CO{-}NH{-}CH(CH_2CH(CH_3)_2){-}COOH + H_2N{-}CH(CH_2C_6H_5){-}CO{-}OCH_3 + C_6H_{11}{-}N{=}C{=}N{-}C_6H_{11}$$

$$\xrightarrow{\text{H-O-N (1-hydroxybenzotriazole)}} (CH_3)_3C{-}O{-}CO{-}NH{-}CH(CH_2CH(CH_3)_2){-}CO{-}NH{-}CH(CH_2C_6H_5){-}CO{-}OCH_3 + C_6H_{11}{-}NH{-}CO{-}NH{-}C_6H_{11}$$

$C_{21}H_{32}N_2O_5$ (392.5)

L-Phenylalanine methyl ester hydrochloride (21.6 g, 100 mmol), 1-hydroxy-benzotriazole monohydrate (15.3 g, 100 mmol), *tert*-butyloxycarbonyl-L-leucine [2] (23.1 g, 100 mmol) and *N*-ethylmorpholine (11.5 g = 12.8 ml, 100 mmol) are dissolved in dry tetrahydrofurane (32 ml), the solution is stirred and cooled in an ice-water bath while dicyclohexylcarbodiimide (21.6 g, 105 mmol) is added. Stirring is continued for one hour at 0 °C and an additional hour at room temperature. The *N*,*N'*-dicyclohexylurea which separated is removed by filtration and the solvent evaporate in vacuo. A mixture of ethyl acetate (500 ml) and of a saturated solution of $NaHCO_3$ in water (250 ml) is added to the residue and the organic phase extracted with a

10% solution of citric acid in water (250 ml), again with saturated $NaHCO_3$ (250 ml) and water (250 ml). The solution is dried over anhydrous Na_2SO_4 filtered [3] and evaporated to dryness in vacuo. The residue is triturated with hexane, filtered, washed with hexane and dried. The crude dipeptide derivative (about 34.4 g, 88%) melts at 80–85°. It is purified by chromatography on a column of basic alumina (100 g) with ethyl acetate as eluent. The pure compound (30.0 g, 76.5%) melts at 91 °C; $[\alpha]_D^{23}$ −27.7° (*c* 1, methanol).

1. König W, Geiger R (1970) Chem Ber 103: 788. 1-Hydroxybenzotriazole is, perhaps, the most frequently used auxiliary nucleophile. For the suppression of racemization, however, 3,4-dihydro-3-hydroxy-1,2,3-benzotriazine-4-one is more effective. Cf. König W, Geiger R (1970) Chem Ber 103: 2024, 2034.
2. Some samples of tert-butyloxycarbonyl-leucine contain one mole of water of crystallization; with such preparations accordingly more has to be used. The presence of the small amount of water added in this way to the reaction mixture does not significantly affect the outcome of the coupling reaction.
3. At this point, some *N,N'*-dicyclohexylurea, still present in the mixture, is removed together with the drying agent.

7.1.3 Coupling with Dicyclohexylcarbodiimide in the Presence of *N*-Hydroxysuccinimide [1]

Phthalyl-L-phenylalanyl-L-valyl-L-glutaminyl-L-tryptophyl-L-leucyl-L-methionyl-L-asparaginyl-*O-tert*-butyl-L-threonine *tert*-butyl Ester [2]

Pht—Phe—Val—Gln—Trp—Leu—OH + Met—Asn—Thr(^{t}Bu)—O^{t}Bu + (cyclohexyl)—N=C=N—(cyclohexyl)

HO—N(CO—CH_2)(CO—CH_2) ⟶ Pht—Phe—Val—Gln—Trp—Leu—Met—Asn—Thr(^{t}Bu)—O^{t}Bu

$C_{65}H_{89}N_{11}O_{14}S$ (1280.6)

Phthalyl-L-phenylalanyl-L-valyl-L-glutaminyl-L-tryptophyl-L-leucine, hemihydrate (8.35 g, 10 mmol), L-methionyl-L-asparaginyl-*O-tert*-butyl-L-threonine *tert*-butyl ester (4.78 g, 10 mmol) and *N*-hydroxysuccinimide (1.15 g, 10 mmol) are dissolved in dimethylformamide (67 ml). The solution is stirred and cooled to −10 °C during the addition of dicyclohexylcarbodiimide (2.06 g, 10 mmol). After two hours at −10 °C and 48 hours at −3 °C water (330 ml) is added [3] to the reaction mixture: the material which separates soon solidifies. It is collected on a filter, thoroughly washed with a saturated solution of $NaHCO_3$ in water (200 ml) then with water (200 ml) and dried over P_2O_5 in vacuo. The crude product (12.1 g) melts with decomposition at 225 to 226 °C. After precipitation from an ethanolic solution with water, the purified nonapeptide derivative (9.7 g, 75%) melts at 226–227 °C; $[\alpha]_D^{20}$ −31.8° (*c* 1, dimethylformamide).

N-Benzyloxycarbonyl-O-tert-butyl-L-tyrosyl-L-leucyl-β-tert-butyl-L-aspartic Acid Methyl Ester [2]

Z–Tyr(^{t}Bu)–OH + H–Leu–Asp(^{t}Bu)–OCH_3·HCl + $N(C_2H_5)_3$ + (cyclohexyl)–N=C=N–(cyclohexyl)

→ (HO–N-succinimide) Z–Tyr(^{t}Bu)–Leu–Asp(^{t}Bu)–OCH_3 + (cyclohexyl)–NH–CO–NH–(cyclohexyl) + HCl·$N(C_2H_5)_3$

$C_{36}H_{51}N_3O_9$ (669.8)

N-Benzyloxycarbonyl-*O*-*tert*-butyl-L-tyrosine (3.72 g, 10 mmol) and L-leucyl-*β*-*tert*-butyl-L-aspartic acid methyl ester hydrochloride (3.54 g, 10 mmol), *N*-hydroxy succinimide (1.20 g, 10.5 mmol) and triethylamine (1.01 g = 1.40 ml, 10 mmol) are dissolved in a mixture dimethylformamide (15 ml) and acetonitrile (15 ml). The solution is cooled to −10 °C and dicyclohexylcarbodiimide (2.06 g, 10 mmol) is added. The reaction mixture is stored at −10° overnight. The precipitate, *N,N'*-dicyclohexylurea, is removed by filtration and the solvents evaporated in vacuo. The residue is dissolved in a mixture of ethyl acetate (100 ml) and water (50 ml) and the organic phase extracted with a 10% solution of citric acid in water (50 ml), 0.5 N $KHCO_3$ (50 ml) and water (50 ml), dried over anhydrous Na_2SO_4 and evaporated to dryness under reduced pressure. Toluene (50 ml) is added and removed in vacuo. The addition and removal of toluene is repeated and the residue is crystallized from diisopropyl ether [4]. The purified fully blocked tripeptide (6.4 g, 96%) melts at 115–116 °C; $[\alpha]_D^{20}$ −20.9° (*c* 2.8, ethanol). On thin layer chromatograms (silica gel) in the system of *n*-heptane-*tert*-butanol-piperidine (3:1:1) only a single spot can be detected.

1. Wünsch E, Drees F (1966) Chem Ber 99: 110
2. Wünsch E, Zwick A, Wendlberger G (1967) Chem Ber 100: 173
3. It seems to be advisable to filter off the separated *N,N'*-dicyclohexyl-urea prior to the addition of water.
4. Diisopropyl ether is a dangerous material which has caused severe explosions.

7.2 Coupling with 1-Isobutyloxycarbonyl-2-isobutyloxy-1,2-dihydroquinoline (IIDQ) [1]

Benzyloxycarbonyl-L-alanyl-L-tyrosine Methyl Ester

C_6H_5–CH_2O–CO–NH–CH(CH_3)–COOH + IIDQ (2-$OCH_2CH(CH_3)_2$, N-CO–$OCH_2CH(CH_3)_2$) →

[C_6H_5–CH_2O–CO–NH–CH(CH_3)–CO–O–CO–$OCH_2CH(CH_3)_2$ + $(CH_3)_2CHCH_2OH$ + quinoline]

→ (H_2N–CH(CH_2–C_6H_4–OH)–CO–OCH_3) C_6H_5–CH_2O–CO–NH–CH(CH_3)–CO–NH–CH(CH_2–C_6H_4–OH)–CO–OCH_3 + CO_2 + $(CH_3)_2CHCH_2OH$

$C_{21}H_{24}N_2O_6$ (400.4)

The coupling reagent, 1-isobutyloxycarbonyl-2-isobutyloxy-1,2-dihydroquinoline [2] (IIDQ, 3.03 g, 10 mmol) is added to a solution of benzyloxycarbonyl-L-alanine (2.23 g, 10 mmol) and L-tyrosine methyl ester [3] (1.95 g, 10 mmol) in dimethylformamide (34 ml) and the mixture stirred at room temperature for about 24 h. The solvent is removed in vacuo, the residue dissolved in ethyl acetate (100 ml) and the solution extracted with a 10% solution of citric acid in water (50 ml), N $NaHCO_3$ (50 ml) and water (50 ml), dried over anhydrous Na_2SO_4 and evaporated to dryness in vacuo. Recrystallization of the residue from ethyl acetate affords the protected dipeptide derivative (3.5 g, 88%) melting at 122–123 °C; $[\alpha]_D^{26} -8.5°$ (c 4, methanol) in analytically pure form.

1. Kiso Y, Kai Y, Yajima H (1973) Chem Pharm Bull 21: 2507. The closely related reagent 1-ethyloxycarbonyl-2-ethyloxy-1,2-dihydroquinoline (Belleau B, Malek G (1968) J Amer Chem Soc 90: 1651 gives lower yields, probably because it can produce second acylation products (urethanes) in somewhat higher amounts than IIDQ.
2. Preparation of the reagent is described on page 198.
3. Fischer E, Schraut W (1907) Justus Liebigs Ann Chem 354: 21; Schwarz H, Bumpus FM (1954) J Amer Chem Soc 81: 890

7.3 The Carbonyldiimidazole Method [1]

Benzyloxy-carbonyl-glycyl-L-tyrosine Ethyl Ester [2]

C_6H_5–CH_2O–CO–NH–CH_2–COOH + Im–CO–Im →

[C_6H_5–CH_2O–CO–NH–CH_2–CO–Im]

$\xrightarrow{H_2N\text{-}CH(CH_2C_6H_4OH)\text{-}CO\text{-}OC_2H_5}$ C_6H_5–CH_2O–CO–NH–CH_2–CO–NH–CH(CH_2–C_6H_4–OH)–CO–OC_2H_5

$C_{21}H_{24}N_2O_6$ (400.4)

Carbonyldiimidazole [3] (1.62 g, 10 mmol) is added to a solution of benzyloxycarbonyl-glycine (2.09 g, 10 mmol) in dry [4] tetrahydrofurane (10 ml). Thirty minutes later [5] L-tyrosine ethyl ester [6] (2.09 g, 10 mmol) is added and the reaction mixture is allowed to stand at room temperature overnight. The solvent is removed in vacuo [7] and N HCl (50 ml) is added to the residue. On cooling a solid separates. This is collected on a filter, washed with water and dried. The crude product (3.93 g, 98%) melts at 125.5–127°.

Recrystallization from 50% ethanol affords the dipeptide derivative in analytically pure form (3.79 g, 95%) with a m.p. of 127–128 °C; $[\alpha]_D^{25}$ +18.2 (*c* 5, abs. ethanol).

1. Staab HA (1957) Justus Liebigs Ann Chem 609: 75
2. Paul R, Anderson GW (1960) J Amer Chem Soc 82: 4596
3. Commercially available. The reagent should be stored and handled with the careful exclusion of moisture. Impure material, e.g. samples contaminated with imidazole hydrochloride, give low yields. Cf. ref. 2.
4. The outcome of the reaction is considerably affected by the presence of moisture; cf. Paul R, Anderson GW (1962) J Org Chem 27: 2094
5. The evolution of CO_2 can be observed. Shorter time between the addition of the reagent and that of the amino component is disadvantageous: the yield is significantly reduced.
6. Fischer E (1901) Ber Dtsch Chem Ges 34: 433. The preparation of this compound is described on page 29.
7. In ref. 2 a stream of air is described as being used for this purpose.

7.4 Coupling with *N*-Ethyl-5-phenylisoxazolium-3′-sulfonate (Woodward's Reagent K) [1]

N^α,N^ε-Dibenzyloxycarbonyl-L-lysyl-glycine Ethyl Ester [2]

C_6H_5–CH_2O–CO–NH–$(CH_2)_4$–CH(NH–CO–OCH_2–C_6H_5)–COOH + $\bar{Cl}\cdot H_3\overset{+}{N}CH_2COOC_2H_5$ + $2N(C_2H_5)_3$ + Woodward's Reagent K (SO_3^-, $\overset{+}{N}$–C_2H_5) ⟶

C_6H_5–CH_2O–CO–NH–$(CH_2)_4$–CH(NH–CO–OCH_2–C_6H_5)–CO–NH–CH_2–CO–OC_2H_5 + $HCl\cdot N(C_2H_5)_3$ + ($SO_3H\cdot N(C_2H_5)_3$)–C_6H_4–CO–CH_2–CO–NH–C_2H_5

$C_{26}H_{33}N_3O_7$ (499.5)

N^α,N^ε-Dibenzyloxycarbonyl-L-lysine [3] (4.14 g, 10 mmol) and *N*-ethyl-5-phenylisoxazolium-3′-sulfonate [4] (2.53 g, 10 mmol) are dissolved in acetonitrile (75 ml), the solution cooled in an ice-water bath and treated with triethylamine (1.01 g = 1.40 ml, 10 mmol) with stirring. Stirring is continued at 0 °C for one hour when a clear solution is obtained. Glycine ethyl ester hydrochloride (1.40 g, 10 mmol) and triethylamine (1.01 g = 1.40 ml, 10 mmol) are added and the reaction mixture is left to stand at room temperature overnight. The solvent is removed under reduced pressure and the residue dissolved in a mixture of ethyl acetate (200 ml) and water (50 ml). The organic layer is washed with a 5% solution of $NaHCO_3$ in water (twice, 50 ml each time), water (50 ml), N HCl (50 ml), water (50 ml), dried over anhydrous Na_2SO_4 and evaporated to dryness in vacuo. The residue (4.91 g, 98%) is crystallized from ether-hexane to yield dibenzyloxycarbonyl-L-lysyl-glycine ethyl ester (4.72 g, 95%) melting at 90.5–92.5 °C [5].

1. Woodward RB, Olofson RA, Mayer H (1961) J Amer Chem Soc 83: 1010
2. Woodward RB, Olofson RA, Mayer H (1966) Tetrahedron 22: Suppl 321
3. Boissonnas RA, Guttmann S, Huguenin RL, Jaquenoud PA, Sandrin E (1958) Helv Chim Acta 41: 1867
4. Commercially available.
5. Repeated recrystallization will raise the m.p. to 91.5–93 °C; Vaughan and Osato (1952) (J Amer Soc 74: 676) report a m.p. of 92–93 °C.

7.5 Coupling with 1-Benzotriazolyl-tri-dimethylaminophosphonium Hexafluorophosphate (BOP-Reagent) [1]

tert-Butyloxycarbonyl-L-threonyl-L-phenylalanine Methyl Ester [2]

$$(CH_3)_3C-O-CO-NH-CH(CHOH-CH_3)-COOH + H_2N-CH(CH_2-C_6H_5)-CO-OCH_3 \xrightarrow{\text{BOP-Reagent}}$$

$$(CH_3)_3C-O-CO-NH-CH(CHOH-CH_3)-CO-NH-CH(CH_2-C_6H_5)-CO-OCH_3$$

$C_{19}H_{28}N_2O_6$ (380.4)

A solution of *tert*-butyloxycarbonyl-L-threonine (2.19 g, 10 mmol) and L-phenylalanine methyl ester hydrochloride (2.16 g, 10 mmol) in acetonitrile (150 ml) is stirred while the reagent [3] (4.42 g, 10 mmol) is added, followed by the addition of triethylamine (2.2 g, 2.8 ml, 20 mmol). Stirring is continued, at room temperature, for one and a half hours. A saturated sodium chloride solution (500 ml) is added and the product extracted with ethyl acetate (three times). The combined extracts are washed with 2N HCl, water, a 5% solution of $NaHCO_3$ in water, and with water. The dried ($MGSO_4$) solution is evaporated in vacuo to dryness. The residue, the blocked dipeptide, (3.74 g, 98%) melts at 94–96 °C, $[\alpha]_D^{22} + 14°$ (*c* 1, EtOAc).

1. Castro B, Evin G, Selve C, Seyer R (1975) Tetrahedron lett 1219
2. Castro B, Dormoy JR, Dourtoglou B, Evin G, Selve C Synthesis 1976: 751
3. The BOP-reagent is commercially available. An improved method for its preparation is presented in ref. 2.

7.6 *O*-Benzotriazolyl-tetramethyluronium Hexafluorophosphate [1]

Preparation of the HBTU reagent [1]

$$(CH_3)_2N-C(=O)-N(CH_3)_2 + COCl_2 \xrightarrow{-CO_2} (CH_3)_2N-C(Cl)=\overset{+}{N}(CH_3)_2\ Cl^- \xrightarrow{NH_4PF_6}$$

$$(CH_3)_2N-C(Cl)=\overset{+}{N}(CH_3)_2\ PF_6^- \xrightarrow[N(C_2H_5)_3]{\text{HO-N (benzotriazole)}} (CH_3)_2N-C(O\text{-benzotriazolyl})=\overset{+}{N}(CH_3)_2\ PF_6^- + (C_2H_5)_3\overset{+}{N}H\ Cl^-$$

$C_{11}H_{16}N_5OBF_6$ (359.1)

These operations must be carried out in a well ventilated hood. A 20% solution of phosgene in toluene (100 ml, 200 mmol) is added dropwise to a solution of tetramethylurea (11.6 g, 100 mmol) in toluene (100 ml). When the evolution of CO_2 ceases (about 15 minutes), anhydrous ether (350 ml) is added with vigorous stirring. The precipitate is collected on a filter, washed with ether (150 ml in three portions) and the hygroscopic material immediately dissolved in chloroform (500 ml). A saturated solution of ammonium hexafluorophosphate (50 ml) is added with vigorous stirring, the organic phase washed with distilled water (40 ml), dried over magnesium sulfate and evaporated in vacuo. The solid residue is washed with ether and dried over phosphorus pentoxide: (24 g, 86%). The (hygroscopic) salt (20 g, 70 mmol) is dissolved in dichloromethane (300 ml) and 1-hydroxybenzotriazole (9.45 g, 70 mmol, cf. footnote 2) is added followed by the addition of triethylamine (9.8 ml, 70 mmol). A white precipitate forms, which is collected on a filter and washed with dichloromethane (100 ml). The product is recrystallized from acetonitrile and dried: 26 g (quantitative yield), m.p. 250 °C [3].

Coupling with HBTU

$$R-COOH + H_2N-R' + HBTU \xrightarrow{N(C_2H_5)_3}$$

$$R-C(=O)-NH-R' + (CH_3)_2N-C(=O)-N(CH_3)_2 + (C_2H_5)_3NHPF_6 + \text{benzotriazolyl-}O^-\ \overset{+}{H}N(C_2H_5)_3$$

The carboxyl-component (10 mmol), the amine-component (10.4 mmol) and triethylamine (20 mmol) are dissolved in acetonitrile (20 ml) and HBTU (3.75 g, 10.4 mmol) is added to the solution. After 15 minutes at room temperature the coupling reaction is complete and the product is isolated in a suitable manner.

1. Dortouglou V, Gross B Synthesis 1984: 572
2. The amount in ref. 1 indicates anhydrous HOBt. The commercially available material is a monohydrate.
3. A melting point of 254 °C was reported by Dortoglou V, Ziegler JC, Gross B, Tetrahedron letters 1978: 1269

IV Removal of Protecting Groups

1 Hydrogenation

1.1 Hydrogenolysis of Benzyl Esters, Benzyl Ethers and of the Benzyloxycarbonyl Group [1]

L-Prolyl-L-leucyl-glycinamide [2]

$$C_6H_5-CH_2O-CO-N(\text{pyrrolidine})-CH-CO-NH-CH(CH_2-CH(CH_3)_2)-CO-NH-CH_2-CO-NH_2 \xrightarrow{H_2/Pd}$$

$$HN(\text{pyrrolidine})-CH-CO-NH-CH(CH_2-CH(CH_3)_2)-CO-NH-CH_2-CO-NH_2 + C_6H_5-CH_3 + CO_2$$

$C_{13}H_{24}N_4O_3$ (284.4)

$C_{13}H_{24}N_4O_3 \cdot 1/2\ H_2O$ (293.4)

A solution of benzyloxycarbonyl-L-prolyl-L-leucyl-glycinamide [3] (41.8 g, 100 mmol) in methanol [4] (250 ml) is prepared in a 500-ml round bottom flask [5] provided with a magnetic stirrer, a gas inlet-outlet tube [6] and surrounded by a large evaporating dish [7]. The air is displaced by a slow stream of nitrogen and a 10% palladium-on-charcoal catalyst [8] (4.2 g = 0.42 g metal) is added. Once again a slow stream of nitrogen is led through the flask, for a few minutes, then the introduction of a slow stream of hydrogen is started. The catalyst is kept in suspension by vigorous stirring. The gas which escapes through the outlet tube [9] is led, from time to time, through a half-saturated filtered solution of $Ba(OH)_2$ in water. When the evolution of CO_2 ceases the reaction mixture is warmed by a water bath of about 50 °C until no more CO_2 can be detected [10] in the escaping gas. The reaction mixture is cooled to room temperature and the introduction of hydrogen is terminated. The remaining gas is displaced by nitrogen and the catalyst is removed by filtration [11], preferably under a blanket of nitrogen. The catalyst is washed with methanol (50 ml) and stored under water until it is discarded or regenerated. The filtrate is evaporated in vacuo [12] or under a stream of air in a well ventilated hood. The residue is dried in air and finally in a desiccator over P_2O_5 in vacuo. The tripeptide amide (hemihydrate) weighs 28.8 g (98%) and melts at 125–126 °C [13].

1. Bergman M, Zervas L (1932) Ber Dtsch Chem Ges 65: 1192
2. This example (M. Bodanszky, unpublished) was selected because the product is obtained in a well-defined form. In most applications of hydrogenolysis salts of the liberated amine are produced and are used as such in the following acylation step. Because of this choice of substrate some comments are necessary on the differences between the conditions applied in this example and the ones conventionally used in the removal of benzyl groups (and the benzyloxycarbonyl group) by hydrogenolysis.
3. Ressler C, du Vigneaud V (1954) J Amer Chem Soc 76: 3107; Zaoral M, Rudinger J (1955) Coll Czech Chem Commun 20: 1183; Boissonnas RA, Guttmann S, Jaquenoud PA, Waller JP (1955) Helv Chim Acta 38: 1491; Rudinger J, Honzl J, Zaoral M (1956) Coll Czech Chem Commun 21: 202
4. It is generally advisable to use ethanol rather than methanol in hydrogenolysis, because the palladium catalyst ignites methanol more readily than other solvents. The solubility of the substrate may necessitate the use of other solvents such as acetic acid, an 8:2 mixture of acetic acid and water, or dimethylformamide.
   If the substrate is soluble in alcohol but the deprotected peptide is not, then water and acetic acid can be added: for instance a mixture of ethanol-water-acetic acid in the volume ratio of 7:2:1 is often useful. The presence of acetic acid or the calculated amount of HCl can be advantageous. In the example described here, no acid was added because the free base was the desired product.
5. There is no obvious advantage in carrying out the reduction under pressure.
6. The inlet tube need not reach below the surface of the solution. It is better to lead the gases just above the surface. The outlet tube is connected through a rubber tubing with a pipette which is immersed into water. This allows a monitoring of the volume of gases which leave the flask. A steady bubbling should be maintained.
7. An empty container is recommended for safety: in case of an accidental breaking of the flask, the reaction mixture is contained and the fire which could start does not spread readily.
8. Other catalysts, such as palladium-black, palladium on $BaSO_4$, etc., do not show special advantages.
9. Because of the escaping hydrogen the operation should be carried out in a well ventilated hood.
10. The escaping gas is led through a pipette into a $Ba(OH)_2$ solution for one minute and the precipitate observed. It is advisable to dip the tube afterwards into N HCl and then into water in order to clean it from $BaCO_3$.
    If acetic acid or 80% acetic acid are used as solvents, the acetic acid carried with the escaping hydrogen gradually dissolves the $BaCO_3$ which formed.
    The end point of CO_2 evolution is usually reached within relatively short time, and generally no heating is necessary. In this case the absence of an acid might be responsible for prolonged CO_2 evolution: the carbamoic acid intermediate is stabilized as a salt by the already formed base. Of course, no CO_2 evolution takes place in the hydrogenolysis of benzyl ethers and benzyl esters.
11. Filtration through a fluted filter paper coated with a mixture of activated charcoal and diatomaceous earth (Celite) is helpful in the removal of the catalyst.
12. The addition of a small amount of water is necessary to allow for the crystallization of the hemihydrate.

13. In the example described here, hydrogenation resulted in a free base. This has the advantage that no tertiary amine has to be added to the reaction mixture in the subsequent acylation step as in the case of amine salt produced by hydrogenolysis in the presence of an acid. Yet, weak non-carboxylic acids, such as 1-hydroxy-benzotriazole, form salts which can be acylated without the addition of tertiary bases (Bodanszky M, Bednarek MA, Bodanszky A (1982) Int J Peptide Protein Res 20: 387).
The cleavage of benzyl ethers and the removal of the benzyl group from the imidazole nucleus of histidine by hydrogenolysis require acidic media. The use of acetic acid or aqueous acetic acid as solvent satisfies this requirement.

1.2 Catalytic Hydrogenation of Methionine Containing Peptides [1, 2]

Benzyloxy-carbonyl-glycyl-L-tryptophyl-L-methionyl-β-*tert*-butyl-L-aspartyl-L-phenylalanine Amide [3]

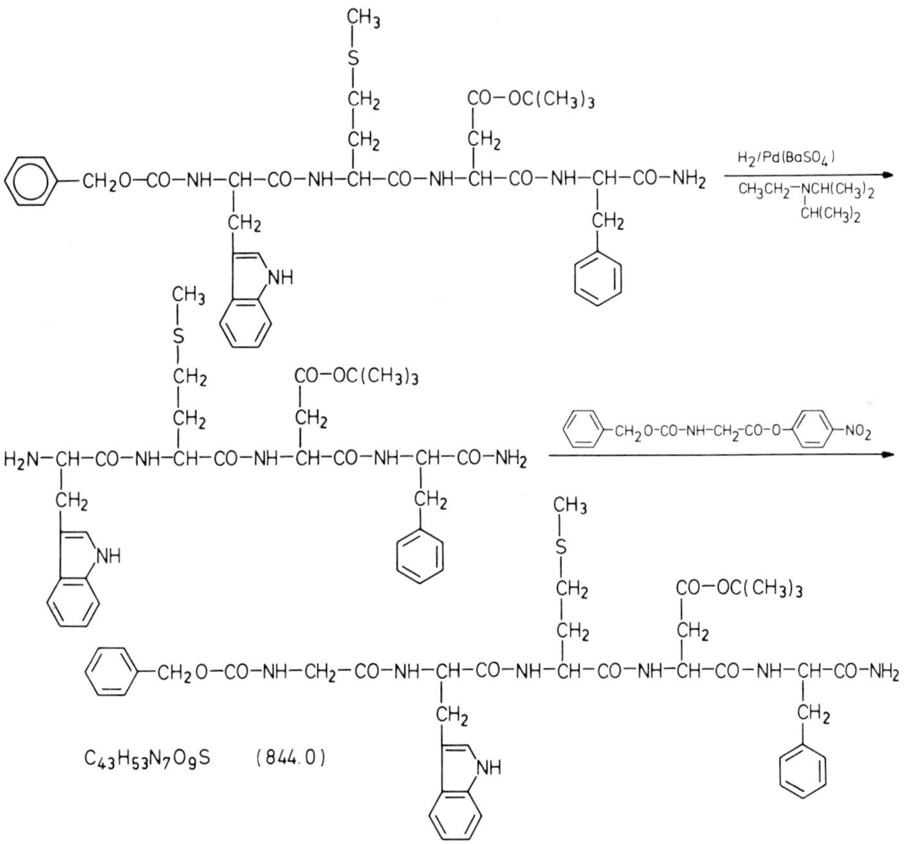
CH₂O–CO–NH–CH–CO–NH–CH–CO–NH–CH–CO–NH–CH–CO–NH₂
H₂/Pd(BaSO₄)
CH₃CH₂–NCH(CH₃)₂ CH(CH₃)₂
H₂N–CH–CO–NH–CH–CO–NH–CH–CO–NH–CH–CO–NH₂
CH₂O–CO–NH–CH₂–CO–O–NO₂
CH₂O–CO–NH–CH₂–CO–NH–CH–CO–NH–CH–CO–NH–CH–CO–NH–CH–CO–NH₂
C₄₃H₅₃N₇O₉S (844.0)

A solution of benzyloxycarbonyl-L-tryptophyl-L-methionyl-β-*tert*-butyl-L-aspartyl-L-phenylalanine amide (7.87 g, 10 mmol) is dissolved in dimethylformamide (450 ml). Distilled water (110 ml) and diisopropylethylamine (16 ml) are added followed by a 10% palladium-on-$BaSO_4$ catalyst (0.80 g).

The mixture is stirred with a magnetic stirrer in an atmosphere of hydrogen [4] until no more gas is absorbed. The catalyst is removed by filtration [5] and the filtrate concentrated in vacuo at a bath temperature of 30 °C to about 80 ml. After dilution with dimethylformamide to about 175 ml, benzyloxycarbonyl-glycine *p*-nitrophenyl ester [6] (3.64 g, 11 mmol) is added. Next day a spot test with ninhydrin indicates complete acylation of the amino group which was set free in the process of catalytic hydrogenation. The solution is diluted with dimethylformamide to about 300 ml and filtered through a column of neutral alumina (60 g); a colorless solution is obtained. The solvent is removed in vacuo and the residue triturated with ether [7]. The solid product is collected on a filter, washed with ether and dried in vacuo. The chromatographically homogeneous [8] pentapeptide derivative (7.85 g, 93%) melts at 188–190 °C; $[\alpha]_D^{25}$ −24.4° (*c* 1.3, dimethylformamide).

1. Medzihradszky-Schweiger H, Medzihradszky K (1966) Acta Chim Acad Sci Hung 50: 539; Medzihradszky-Schweiger H (1973) ibid 76: 437
2. Catalytic reduction in the presence of a strong base allows the removal of benzyloxycarbonyl groups and hydrogenolysis of benzyl esters of methionine containing peptides.
   Peptides with more than one methionine residue might be less suitable substrates (cf. ref. 3), while the presence of *S*-benzyl-cysteine in the sequence prevents the application of the method. On the other hand, the benzyloxycarbonyl group can be removed from *S*-alkyl-cysteine containing peptides by hydrogenation in liquid ammonia (Kuromizu K, Meienhofer J (1974) J Amer Chem Soc 96: 4978; Felix AM, Jimenez MH, Mowles T (1978) Int J Peptide Protein Res 11: 829), particularly in the presence of dimethylacetamide and triethylamine (Felix AM, Jimenez MH, Meienhofer J (1980) Org. Synth 59: 159)
3. Bodanszky M, Martinez J, Priestley GP, Gardner JD, Mutt V (1978) J Med Chem 21: 1030
4. A closed system is used. The gas burette and niveau-vessel are filled with a mixture of dimethylformamide-water-diisopropylethylamine in the ratio used in the reaction mixture. First the air is displaced by nitrogen then the latter by hydrogen. The uptake of hydrogen is recorded from time to time. It stops when the calculated amount is absorbed. The hydrogen is replaced by nitrogen before the reaction vessel is opened.
5. Catalytic hydrogenation in dimethylformamide with various palladium catalysts usually results in dark colored solutions. The colored product, probably a colloid, is removed by filtration through a column of neutral alumina. This is more conveniently done after the next operation, the acylation of the amino group.
6. Bodanszky M (1957) Acta Chim Acad Sci Hung 10: 335. Preparation of *p*-nitrophenyl esters is described in this volume; cf. page 97
7. Peroxide free ether should be used.
8. A single spot appears on thin layer chromatograms (silica gel) with R_f 0.85 in the system of *n*-butanol-acetic acid-water (4:1:1) and R_f 0.6 in the system chloroformmethanol (9:1).

1.3 Transfer Hydrogenation [1, 2] with 1,4-Cyclohexadiene as Hydrogen Donor [3, 4]

N^{ε}-tert-Butyloxycarbonyl-L-lysyl-*O-tert*-butyl-L-threonine Methyl Ester

```
(CH3)3C-O-CO-NH           CH3
              |            |
            (CH2)4        CH-OC(CH3)3
              |            |                      Pd
C6H5-CH2O-CO-NH-CH-CO-NH-CH-CO-OCH3   ------------->
                                         (cyclohexadiene)

(CH3)3C-O-CO-NH           CH3
              |            |
            (CH2)4        CH-OC(CH3)3
              |            |
         H2N-CH-CO-NH-CH-CO-OCH3   +  CO2  +  C6H5-CH3
```

$C_{20}H_{39}N_3O_6$ (417.5)

A solution of N^{α}-benzyloxycarbonyl-N^{ε}-*tert*- butyloxycarbonyl-L-lysyl-*O-tert*-butyl-L-threonine methyl ester (5.52 g, 10 mmol) is dissolved in absolute ethanol (50 ml) [5]. The solution is placed in a round bottom flask provided with a magnetic stirrer [6], gas inlet-outlet tubes and surrounded by a water bath of 25 °C. A slow stream of nitrogen is lead above the surface of the solution and a 10% palladium-on-charcoal catalyst (5.5 g) [7] is added with vigorous stirring, followed by 1,4-cyclohexadiene (8.0 g = 9.4 ml, 100 mmol). Stirring is continued and the progress of the reaction is monitored by thin layer chromatograms. At the completion of the hydrogenolysis the catalyst is removed by filtration [8] and the solvent evaporated under reduced pressure. The residue (about 4.1 g), the partially deprotected dipeptide derivative, is dried in vacuo.

1. Anantharamaiah GM, Sivanandaiah KM, J Chem Soc Perkin I 1977: 490
2. Jackson AE, Johnstone RAW, Synthesis 1977: 685
3. Felix AM, Heimer EP, Lambros TJ, Tzougraki C, Meienhofer J (1978) J Org Chem 43: 4194
4. Cyclohexadiene is a more effective hydrogen donor than the previously recommended (ref. 1, 2) cyclohexene. Several other donors such as formic acid, hydrazine are potentially useful but could react with the peptide derivative subjected to hydrogenolysis or with the product. Transfer hydrogenation is suitable for the removal of benzyloxycarbonyl groups, for the hydrogenolysis of benzyl esters and benzyl ethers. Reduction of nitroguanidine groups and removal of the benzyl group from the imidazole nucleus of histidine requires an efficient catalyst, that is palladium black rather than palladium-on-charcoal. In glacial acetic acid as solvent the process is applicable for methionine containing peptides as well, but is not satisfactory with compounds which contain *S*-benzyl cysteine residues.
5. Transfer hydrogenation is faster in glacial acetic acid than in ethanol. The reaction rate is further reduced in methanol, dimethylacetamide and dimethylformamide and is impractically low in isopropanol, trifluoroethanol, trifluoroacetic acid, phenol, hexamethylphosphoramide and dimethylsulfoxide. Concentrations ranging for 0.05 to 0.25 molar are optimal.
6. An efficient execution of hydrogenation is possible in a cylindrical vessel (e.g. 3 × 25 cm) provided with an adapter which carries gas inlet and outlet tubes and a vibro-mixer. This apparatus is particularly useful when palladium black is the catalyst because magnetic stirring might not be vigorous enough to disperse the catalyst and the hydrogen sufficiently.

7. If more than one benzyl group has to be removed a higher amount of catalyst might be necessary. For some peptides the amounts needed could turn out to be impractical. In such cases palladium black (10% of the substrate for each protecting group) might be useful.
8. It is advisable to lead a slow stream of argon or nitrogen above the filter. The use of filter-aid (diatomaceous earth, such as Celite), greatly facilitates the operation.

2 Reduction with Sodium in Liquid Ammonia [1]

L-Glutaminyl-L-asparagine [2]

$$CH_3-C_6H_4-SO_2-NH-CH(CH_2-CH_2-CONH_2)-CO-NH-CH(CH_2-CONH_2)-COOH \xrightarrow{Na/NH_3} H_2N-CH(CH_2-CH_2-CONH_2)-CO-NH-CH(CH_2CONH_2)-COOH$$

$C_9H_{16}N_4O_5$ (260.3)

A sample of *p*-toluenesulfonyl-L-glutaminyl-L-asparagine [2] is dried in a desiccator, over P_2O_5, overnight in vacuo. The dried protected dipeptide acid (4.14 g, 10 mmol) is added with stirring to liquid ammonia (about 300 ml) [3]. Stirring is continued and clean sodium is added in small pieces [4] until the blue color, which develops on dissolution of the metal and disappears during the reaction, persists for about one minute. Approximately 1.45 g sodium is required to reach this point. Glacial acetic acid (5 ml) is added [5] and the ammonia is allowed to evaporate. To remove the residual ammonia [6] the flask is placed in a desiccator and evacuated with the help of a water aspirator for at least one hour. The residue is dissolved in distilled water (40 ml), the pH is adjusted to 6 with glacial acetic acid and the solution diluted with absolute ethanol (800 ml). Scratching of the walls with a glass rod induces crystallization of the free dipeptide. After overnight storage in the refrigerator, the crystals are collected on a filter, washed with absolute ethanol (50 ml) and dried in air. The product (about 2.1 g, 80%) melts with decomposition at 200–203 °C. Recrystallization from water-ethanol yields purified L-glutaminyl-L-asparagine melting at 210–211 °C dec; $[\alpha]_D^{21}$ +20.8° (*c* 2.7, 0.5 N HCl). On elemental analysis correct values are obtained for C, H and N.

1. du Vigneaud V, Behrens OK (1937) J Biol Chem 117: 27. In addition to cleaving the *p*-toluenesulfonyl group, reduction with sodium in liquid ammonia removes the benzyloxycarbonyl groups and *O*-benzyl groups as well.
2. Swan JM, du Vigneaud V (1954) J Amer Chem Soc 76: 3110
3. Liquid ammonia is collected in a 500 ml round bottom flask provided with a magnetic stirrer and with gas inlet and outlet tubes. The flask is surrounded by a large crystallizing dish containing acetone. Dry NH_3 gas is passed through the vessel to displace air and moisture and then cooling is started by the addition of dry ice to the acetone. The NH_3 stream is increased to maintain a slight positive pressure in the flask. The stream can be best regulated if the escaping ammonia is led through a bubbler filled with silicone oil. During the entire operation some ammonia is allowed to escape. Of course, *the reaction is carried out in a well*

ventilated hood. When the desired amount of liquid condensed the cooling bath is removed and replaced by an empty dish of similar size. The ammonia stream is gradually decreased and completely stopped when the liquid reached its boiling point: the escaping ammonia blocks the entry of moist air and provides the necessary anhydrous conditions. The absence of water can be checked by the addition of a small piece (a few mg) of sodium: the blue color should persist for a minute or longer. The frost which deposits on the flask should not be removed, since it provides the needed insulation, but a small window can be made by wiping off the ice at one point with cotton moistened with ethanol. The flask is kept open during the addition of the reactants and closed afterwards with a cotton-filled drying tube.

4. The sodium is "shaved" with a sharp knife, cut into small pieces, weighed under hexane and kept under hexane until used. Unused sodium and the shavings should be disposed of by adding them carefully to ethanol.
5. There is a strong exothermic reaction between acetic acid and ammonia. The liquid boils vigorously and some noise is generated. It is advisable to let the acid run down the wall of the flask. A wide mouth pipette should be used to prevent clogging by ammonium acetate. For larger quantities instead of acetic acid, ammonium acetate can be applied.
6. The seemingly dry residue might contain enough ammonia to cause serious problems if inhaled.

3 Removal of the Phthalyl (Phthaloyl) Group by Hydrazinolysis [1]

Glycyl-glycine

Phthalyl-N$-CH_2-CO-NH-CH_2-COOH$ + H_2NNH_2 ⟶ phthalhydrazide enolate $^-O \cdot \overset{+}{H_3N}-CH_2-CO-NH-CH_2-COOH$

$\xrightarrow{HCl}$ phthalylhydrazine + $HCl \cdot H_2N-CH_2-CO-NH-CH_2-COOH$ $\xrightarrow{OH^-}$

$H_2N-CH_2-CO-NH-CH_2-COOH$

$C_4H_8N_2O_3$ (132.1)

Phthalyl-glycyl-glycine (2.62 g, 10 mmol) is added to a 1 M solution of hydrazine hydrate [2] in absolute ethanol (10 ml). The mixture is diluted with ethanol (30 ml) and heated under reflux for one hour [3]. The alcohol is removed in vacuo, the residue treated with 2 N HCl (25 ml) at 50 °C for 10 minutes [4] and kept at room temperature for 30 min. The insoluble phthalylhydrazine is removed by filtration and the filtrate evaporated to dryness in vacuo. The dipeptide hydrochloride is recrystallized from boiling ethanol to afford the purified salt (monohydrate, 1.73 g) which is then dissolved in water (20 ml) and treated with the anion exchange resin [5] Amberlite-IR4B until a drop of the solution gives no positive reaction for chloride ion with $AgNO_3-HNO_3$. The resin is filtered off, washed with water and the filtrate concentrated to about 20 ml. The solution is heated on a stream bath, diluted with absolute ethanol until crystallization starts and allowed to cool to room temperature. The crystals, colorless plates, are collected, washed with 95% ethanol and dried. The pure dipeptide thus obtained (1.07 g, 81%) decomposes, without melting, at 215–222 °C.

1. Sheehan JC, Frank VS (1949) J Amer Chem Soc 71: 1856; cf. also King FE, Kidd DAA, J Chem Soc 1949: 3315
2. Sensitive peptides can be deprotected under milder conditions, e.g. with a 2 M solution of hydrazine acetate at 50 °C for two hours (Schwyzer R, Costopanagiotis A, Sieber P (1963) Helv Chim Acta 46: 870)
3. The phthalyl group is removed by hydrazine also at room temperature in one or two days.
4. Instead of heating for 10 minutes at 50 °C stirring at room temperature for about 2 hours will also convert the enolate salt to phthalylhydrazine and the dipeptide to the hydrochloride.
5. The ion exchange resin can be used either in OH or in acetate cycle. The commercially available resin is usually the chloride. It is converted to the acetate by repeated treatments with a 2 M solution of sodium acetate in water followed by thorough washing with distilled water.

4 Acidolysis

4.1 Hydrobromic Acid in Acetic Acid [1]

Glycyl-L-phenylalanine Benzyl Ester Hydrobromide [2]

$C_6H_5-CH_2O-CO-NH-CH_2-CO-NH-CH(CH_2C_6H_5)-CO-OCH_2-C_6H_5 \xrightarrow[\text{20 min room temp.}]{\text{HBr/AcOH}}$

$HBr \cdot H_2N-CH_2-CO-NH-CH(CH_2C_6H_5)-CO-OCH_2-C_6H_5 + C_6H_5-CH_2Br + CO_2$

$C_{18}H_{21}N_2O_3Br$ (393.3)

An about 33% solution of HBr in acetic acid [3] (10 ml) is placed in a 250 ml round bottom flask and benzyloxycarbonyl-glycyl-L-phenylalanine benzyl ester [2] (4.46 g, 10 mmol) is added with swirling. The flask is closed with a cotton-filled drying tube and swirled to effect complete dissolution of the protected dipeptide. Vigorous evolution of CO_2 takes place [4] and some evolution of heat can be observed [5]. When the gas evolution ceases (about 20 minutes) [6] dry ether (200 ml) is added with swirling and the reaction mixture is stored in a refrigerator [7] for several hours. The precipitate is collected on a filter, washed with ether (50 ml) and dried over NaOH pellets in vacuo. Glycyl-L-phenylalanine benzyl ester hydrobromide (3.4 g, 87%) can be purified by recrystallization from ethanol-ether and it melts then at 193 °C; correct analytical values are obtained for N and Br.

***S*-Benzyl-L-cysteinyl-L-prolyl-L-leucyl-glycinamide [8]**

$C_6H_5-CH_2O-CO-NH-CH(CH_2-S-CH_2-C_6H_5)-CO-N(Pro)-CH-CO-NH-CH(CH_2-CH(CH_3)_2)-CO-NH-CH_2-CO-NH_2 \xrightarrow[\text{2. Amberlite IRA-400}]{\text{1. HBr/AcOH}}$

$H_2N-CH(CH_2-S-CH_2-C_6H_5)-CO-N(Pro)-CH-CO-NH-CH(CH_2-CH(CH_3)_2)-CO-NH-CH_2-CO-NH_2$

$C_{23}H_{35}N_5O_4S$ (477.6)

N-Benzyloxycarbonyl-*S*-benzyl-L-cysteinyl-L-prolyl-L-leucyl-glycinamide [9] (6.12 g, 10 mmol) is suspended [10] in glacial acetic acid (12.5 ml) [11] in a 250 ml round bottom flask. An about 4 molar solution of HBr in acetic acid [3] (25 ml) [11] is added with swirling and the flask is closed with a cotton-filled drying tube. Swirling is continued until the entire amount of the protected peptide is dissolved and then the mixture allowed to stand at room temperature for one hour. Dry ether (200 ml) is added with swirling, the precipitate collected on a filter [12], washed with ether (100 ml) and dried in a desiccator over NaOH pellets in vacuo. The dried hydrobromide [13] is dissolved in methanol (80 ml) and an anion exchange resin, such as Amberlite IRA-400 (in OH cycle) is added in small portions until a drop of the solution gives no precipitate with $AgNO_3$ in dilute nitric acid. The resin is removed by filtration and thoroughly washed with methanol. The filtrate and washings are combined and the solvent removed under reduced pressure. The residue (a free base) is a semisolid mass (5.3 g) [14] than can be used in the subsequent acylation step [15].

1. Ben-Ishai D, Berger A (1952) J Org Chem 17: 1564
2. Ben-Ishai D (1954) J Org Chem 19: 62
3. For the preparation of the HBr/AcOH reagent acetic acid should be distilled. A small fore-run is discarded and about 10% is left in the distillation flask. Dry HBr gas is led from a cylinder into a weighed amount of the distilled acetic acid and after a few minutes the recipient vessel (which is protected from moisture by a cotton filled drying tube) is cooled with ice water. After about one hour of gas introduction the weight-increase is determined. The solution should be stored in glass bottles with glass stoppers. The ground surfaces and the bottles themselves should be cleaned with acetic acid. (The fore-run can be used for this purpose). Solutions of HBr in acetic acid prepared in this way are colorless or at most straw colored and remain light colored for several months if stored in the refrigerator. It is advisable to use bottles which in addition to ground glass stoppers have a glass cap as well.
4. The escaping gas contains some HBr. The entire operation should be performed in a well ventilated hood.
5. At higher temperature a major loss in the benzyl ester group will occur. Thus, with larger amounts the reaction mixture should be cooled to about 20 °C.
6. The benzyloxycarbonyl group is preferentially removed but the benzyl ester is also cleaved by HBr/AcOH except at a lower rate. Therefore, the reaction time must be kept within the limits indicated.
7. An explosion-proof refrigerator should be used.
8. Bodanszky M, du Vigneaud V (1959) J Amer Chem Soc 81: 5688
9. Bodanszky M, du Vigneaud V (1959) J Amer Chem Soc 81: 2504
10. Direct contact between the dry powdered peptide derivative and the HBr/AcOH reagent often results in the formation of lumps which are not readily dissolved. Dissolution of the protected peptide in the reagent is greatly facilitated if the material is suspended (or dissolved, if sufficiently soluble) in acetic acid. In order to achieve ready cleavage of the benzyloxycarbonyl group the concentration of HBr in the reaction mixture should be not less than 2 molar.
11. A certain economy in the use of acetic acid and the HBr/AcOH reagent is recommended: a disproportionately larger amount of ether is needed for precipitation from dilute solutions.
12. The peptide hydrobromides are usually hygroscopic, at least in their crude form. Therefore, a sinter glass filter should be used and kept loosely covered during filtration. The precipitate must be kept covered with ether and after completion of the washing the filter with the

product should be immediately transferred into a desiccator and dried over NaOH pellets in vacuo. It is often practical to dissolve the hydrobromide on the filter with small portions of dimethylformamide, to filter the solution into a round bottom flask and to use this solution, after the addition of an appropriate amount of tertiary base, in the subsequent acylation step.
It should be remembered that benzyl bromide is formed in the here described acidolysis. The filtrates which contain in addition to this potent lachrimator also HBr, acetic acid and ether should be disposed of accordingly. With smaller quantities the ether can be allowed to evaporate under a hood and ammonia added in excess to the solution. On standing benzyl bromide will be converted to benzylamine and this will cause fewer problems.

13. Generally a monohydrobromide is not obtained at this point. The amide nitrogens are sufficiently basic to form salts with HBr, at least under the conditions of the acidolysis described above. On prolonged storage in vacuo over NaOH pellets or better on recrystallization from methanol-ether, the hydrobromides are secured in pure, non-hygroscopic form, without excess HBr.
14. This is more than the calculated amount (4.48 g). Obviously some solvent is retained in the material. The tetrapeptide amide can be crystallized from water (Ressler C, du Vigneaud V (1954) J Amer Chem Soc 76: 3107). The crystals melt at 69–71 °C; $[\alpha]_D^{23}$ −47.7° (*c* 1, absolute ethanol). Analysis indicates 1.5 moles of water of crystallization. Prolonged drying over P_2O_5 in vacuo removes the water from the crystals.
15. Acylation of this tetrapeptide amide with benzyloxycarbonyl-L-asparagine *p*-nitrophenyl ester is described on p. 109

4.2 Hydrobromic Acid in Trifluoroacetic Acid [1, 2]

α-Melanotropin [1]

N–Ac–Ser–Tyr–Ser–Met–Glu(OBzl)–His–Phe–Arg–Trp–Gly–Lys(Z)–Pro–Val–NH2 $\xrightarrow[CF_3COOH]{HBr}$

N–Ac–Ser–Tyr–Ser–Met–Glu–His–Phe–Arg–Trp–Gly–Lys–Pro–Val–NH2 + CO_2 + C_6H_5–CH_2Br
(hydrobromide)

N-Acetyl-L-seryl-L-tyrosyl-L-seryl-L-methionyl-γ-benzyl-L-glutamyl-L-histidyl-L-phenylalanyl-L-arginyl-L-tryptophyl-glycyl-N^{ε}-benzyloxycarbonyl-L-lysyl-L-prolyl-L-valinamide [1] (385 mg, 0.20 mmol) is dissolved in trifluoroacetic acid (20 ml); diethyl phosphite [3] (5 ml) and ethyl methyl sulfide [3] (5 ml) are added and the solution is cooled to −5 °C while protected from light. A vigorous stream of dry HBr gas is introduced into the solution for one hour. The flask is closed with a glass stopper and kept at 20 °C, protected from light, for two hours. The solvent and the volatile materials are removed in vacuo with the help of a rotary evaporator at a bath temperature of 30 °C and the residue is triturated with ether (200 ml) [4] until a powder is formed. The product is collected by centrifugation, resuspended in ether and centrifuged. Washing with ether in this way is repeated twice more. The material is dried in vacuo then dissolved in distilled water (60 ml) and the pH adjusted to 4.0 with 4 N NH_4OH. The dissolved ether is removed by a stream of nitrogen [5] led through the solution. Bioassay indicates a total of 775000 units of α-melanotropin. The hormone is isolated by countercurrent distribution and further purified by electrophoresis [6].

1. Guttmann S, Boissonnas RA (1959) Helv Chim Acta 42: 1257
2. Deprotection of serine containing peptides with HBr in acetic acid results in extensive *O*-acetylation. Threonine residues are acetylated to a lesser extent. No *O*-acylation occurs in HBr-trifluoroacetic acid. Alternatively, in order to avoid *O*-acylation, HBr can be introduced into a solution of the protected peptide in a 1:1 mixture of phenol and *p*-cresol (Bodanszky M, Tolle JC, Deshmane SS, Bodanszky A (1978) Int J Peptide Protein Res 12: 57)
3. Alkylation of the thioether in methionine and of the aromatic nuclei of tyrosine and tryptophan by benzyl bormide is prevented by the addition of diethyl phosphite and ethyl methyl sulfide. These compounds are volatile and toxic and have unpleasant odors as well. Thus, the entire operation must be carried out in a well ventilated hood. The use of HBr and trifluoroacetic acid requires the same precaution.
4. Peroxide-free ether should be used.
5. In the original publication (ref. 1), hydrogen is described.
6. The procedures of isolation and purification are described in detail in ref. 1.

4.3 Hydrochloric Acid in Acetic Acid [1]

Nitro-L-arginyl-copolymer [1]

```
                 NHNO2
                 ||
              HN-C-NH2
                 |
 CH3           (CH2)3                      |
 |               |                         CH2
CH3-C-O-CO-NH-CH-CO-OCH2-C6H4-CH        --HCl/AcOH-->
 |                                         CH2
 CH3                                       |

                                      NHNO2
                                      ||
                                   HN-C-NH2
                                      |
                                    (CH2)3                   |
                                      |                      CH2
(CH3)3CCl + CO2 + HCl·H2N-CH-CO-OCH2-C6H4-CH      --N(C2H5)3-->
                                                             CH2
                                                             |

               NHNO2
               ||
            HN-C-NH2
               |
             (CH2)3                  |
               |                     CH2
         H2N-CH-CO-OCH2-C6H4-CH
                                     CH2
                                     |
```

A sample of *tert*-butyloxycarbonyl-nitro-L-arginyl-polymer [2] (10 g, 1.74 mmol nitro-L-arginine content) is placed into a reaction vessel provided with a sinter glass bottom [3] and a one molar solution of HCl in acetic acid (30 ml) is added. The suspension is shaken at room temperature for 30 minutes. The solution is filtered through the sinter glass bottom and the resin remaining in the vessel thoroughly washed [4] with acetic acid (3 times), ethanol (3 times) and dimethylformamide [5] (3 times). The remaining hydrochloride of the nitro-L-arginyl-polymer is treated with dimethylformamide containing 3 ml triethylamine for ten minutes then washed with dimethylformamide (3 times, 3 minutes each time). The aminoacyl polymer is now ready for the incorporation of the next amino acid residue.

1. Merrifield RB (1964) Biochemistry 3: 1385
2. Copolystyrene-2%-divinylbenzene beads, 200–400 mesh.
3. The details of the vessel are described by Merrifield (1963) (J Am Chem Soc 85: 2149)
4. About 30 ml of solvent is needed for each wash and the solvent should be kept in contact with the resin at least for 3 minutes. It is important to completely remove all acetic acid from the resin, because otherwise it will cause acetylation of the amino group in the subsequent carbodiimide mediated coupling.
5. Purified with barium oxide (Thomas AB, Rochow EG (1957) J Am Chem Soc 79: 1843).

4.4 Trifluoroacetic Acid [1, 2]

L-Methionyl-L-glutaminyl-L-histidyl-L-phenylalanyl-L-arginyl-L-tryptophyl-glycinamide Acetate [1]

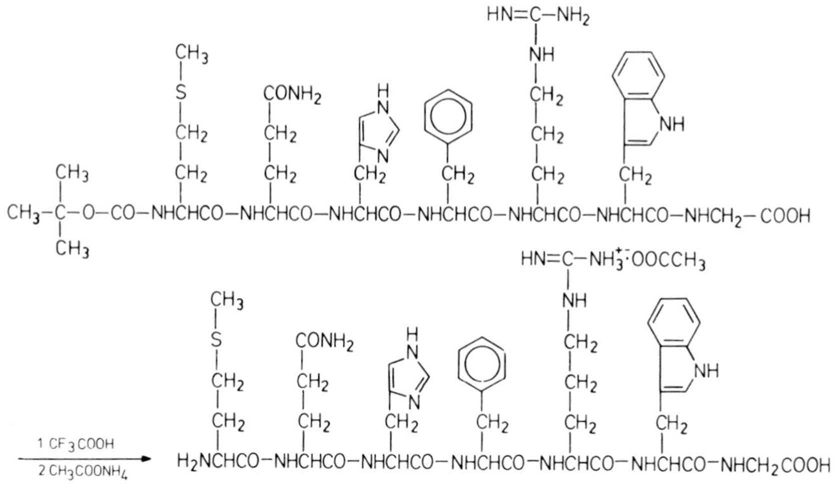

$(CH_3)_3C{-}O{-}CO{-}NHCHCO{-}NHCHCO{-}NHCHCO{-}NHCHCO{-}NHCHCO{-}NHCHCO{-}NHCH_2{-}COOH$

1 CF_3COOH
2 CH_3COONH_4

$H_2NCHCO{-}NHCHCO{-}NHCHCO{-}NHCHCO{-}NHCHCO{-}NHCHCO{-}NHCH_2COOH$ (arginine side chain: $HN{=}C{-}NH_3^+ \cdot {}^-OOCCH_3$)

$C_{46}H_{64}N_{14}O_{11}S$ (1021.1)

tert-Butyloxycarbonyl-L-methionyl-L-glutaminyl-L-histidyl-L-phenylalanyl-L-arginyl-L-tryptophyl-glycine [1] (1.1 g, 1.0 mmol) is dissolved in trifluoroacetic acid (8.5 ml) and left to stand [3] at room temperature for one hour [4]. The solution [5] is evaporated [6] to dryness in vacuo [7] and the residue triturated with dry ether [8]. The solid product is disintegrated under fresh ether, filtered, washed with ether and dried in vacuo. The crude material [9] (1.33 g, calculated for tri-trifluoroacetate [10] 1.40 g) is purified by countercurrent distribution between *n*-butanol and a 0.3 molar ammonium acetate buffer of pH 7.1 through 70 transfers (20 ml phases). The heptapeptide acetate travels with a K value of 0.55 and is recovered from tubes No. 20 through 28. The solvent is removed by evaporation in vacuo, ammonium acetate by sublimation at 40 °C in high vacuum. The residue (0.67 g) is crystallized from ethanol-water to afford the pure heptapeptide (436 mg, 42%) as the mono-acetate; $[\alpha]_D^{26}$ -25.3° (*c* 1, dimethylformamide). Recrystallization from ethanol-water renders the product analytically pure.

1. Kappeler H, Schwyzer R (1960) Helv Chim Acta 43: 1453
2. Trifluoroacetic acid readily cleaves the *tert*-butyloxycarbonyl group. *tert*-Butyl esters and *tert*-butyl ethers require longer reaction times. The reagent is relatively inert toward the benzyloxycarbonyl group, benzyl ester and benzyl ethers, but a slight loss of benzyloxycarbonyl groups usually accompanies the cleavage of *tert*-butyloxycarbonyl groups. The *O*-benzyl group is gradually removed from the tyrosine side chain and in part it migrates to the ortho position of the aromatic ring with the formation of 3-benzyl-tyrosine residues. Selectivity is improved if instead of benzyl derivatives the more acid resistant negatively substituted benzyl derivatives are applied; these necessitate, however, the use of very strong acids in the step of final deprotection.
   The importance of trifluoroacetic acid lies in its general usefulness as solvent for almost all peptides and peptide derivatives. Therefore, other, more selective acidic reagents are soldom used. The selectivity of acidolysis with trifluoroacetic acid is improved by dilution with water (Schnabel E, Klostermeyer H, Berndt H (1971) Justus Liebigs Ann Chem 749: 90), with acetic acid (Klausner YS, Bodanszky M (1973) Bioorg Chem 2: 354) or with phenols (Bodanszky M, Bodanszky A, Int J Peptide Protein Res (1984) 23: 565.
3. From timė to time the glass stopper is lifted to allow the escape of CO_2. For larger quantities the reaction vessel is provided with a $CaCl_2$-filled drying tube.
4. With peptides carrying benzyl and/or benzyloxycarbonyl groups higher temperature should be avoided if the benzyl-based blocking groups are to be retained. Even at room temperature longer exposure to trifluoroacetic acid removes the benzyloxycarbonyl group and also cleaves benzyl ethers on tyrosine side chains. Some loss occurs even within the hour described in the here cited example. Removal of the *tert*-butyloxycarbonyl group with neat trifluoroacetic acid is generally complete in 15 minutes (at room temperature).
5. Some decomposition of the indole nucleus is caused by trifluoroacetic acid and the solution turns bluish-purple. This damage to tryptophan residues can be kept at a minimum by cooling the reaction mixture in an ice-water bath and by displacing the air with argon or nitrogen.
6. Precipitation (with ether) instead of evaporation would diminish the effect of *tert*-butyl trifluoroacetate formed in the reaction (Lund BF, Johansen NL, Vølund A, Markussen J (1978) Int J Peptide Protein Res 12: 258). Yet, the presence of excess trifluoroacetic acid renders peptides more soluble in ether and serious losses can occur. Also, such precipitation often results in milky emulsions from which the product can be recovered only with some difficulty.
7. A water aspirator is sufficient for the removal of most of the trifluoroacetic acid. The rest can be removed with the help of an oil pump, protected with appropriately cooled traps.
8. Peroxide-free ether should be used: tryptophan containing peptides are sensitive to oxidation, particularly under acidic conditions and the methionine side chain is readily oxidized to a sulfoxide by peroxides.
9. In numerous other examples the crude trifluoroacetate is used (in the presence of a tertiary base) in the subsequent acylation step. In order to avoid trifluoroacetylation *p*-toluenesulfonic acid can be added, in the calculated amount, to the reaction mixture before evaporation and the *p*-toluenesulfonate salt of the peptide isolated by trituration with ether.
10. The weakly basic amide bonds can form salts with strong acids and a product with more than the calculated amount of trifluoroacetic acid is obtained. The loosely bound acid is removed, however, on prolonged drying in vacuo or by reprecipitation from a solution (in alcohol or dimethylformamide) with ether.

4.5 Removal of Protecting Groups with Trifluoroacetic Acid-Thioanisole [1]

Porcine Vasoactive Intestinal Peptide [2]

Z(OMe)–His–Ser(Bzl)–Asp–Ala–Val–Phe–Thr–Asp–Asn–Tyr–Thr–Arg(Mts)–Leu–Arg(Mts)–

Lys(Z)–Gln–Met(O)–Ala–Val–Lys(Z)–Lys(Z)–Tyr–Leu–Asn–Ser–Ile–Leu–As–NH_2 [3, 4]

1 HS–C_6H_5; 2 TFA/ C_6H_5–SCH_3 ⟶ porcine VIP

1. Reduction of the methionine sulfoxide residue to a methionine residue. A sample of the protected 28-peptide [5] (515 mg, 0.12 mmol) is dissolved in dimethylformamide (5 ml), thiophenol (1.3 g, 1.2 ml, 12 mmol) is added and the mixture kept at 60 °C for one day. The solvent is removed in vacuo and the residue triturated with ethyl acetate until a solid forms. The reduced material is collected by filtration, washed with ethyl acetate and dried in vacuo over P_2O_5 at room temperature for 3 hours: 0.48 g (94%).

2. Complete deprotection. A sample of the reduced intermediate (57 mg, 13 μmol) is dissolved in a mixture of trifluoroacetic acid (6 ml) and thioanisole (0.48 g, 0.46 ml, 3.9 mmol) and *m*-cresol (0.43 g, 0.41 ml, 3.9 mmol). After about 10 hours at 28° the mixture is diluted with ether until a precipitate forms. The solid material is collected on a filter, redissolved in a similar mixture of trifluoroacetic acid, thioanisole and *m*-cresol and stored at 28 °C for about 14 hours. The free peptide (trifluoroacetate salt) is precipitated with ether, dissolved, in distilled water and treated with an anion exchange resin [6] in acetate form (about 2 g). After about 30 min the suspension is filtered and the resin washed (slowly) with water. The pH of the filtrate is adjusted to 9 with 5% NH_4OH and 30 min later to pH 6 with N acetic acid. Water and ammonium salts are removed by lyophilization and the peptide purified by chromatography on a Sephadex G-25 column (1.5 × 130 cm) with 0.1 N acetic acid as eluent. Fractions of 5 ml are collected and the elution is monitored by uv absorption at 275 nm. The purified peptide is recovered from fractions 41–51. Homogeneous material (22 mg, 43%) is secured by a second chromatography on an ion exchange column [7–10].

1. Kiso Y, Ukawa K, Nakamura S, Ito K, Akita T (1980) Chem Pharm Bull 28: 673
2. Takeyama M, Koyama K, Inoue K, Kawano T, Adachi T, Tobe T, Yajima H (1980) Chem Pharm Bull 28:1873
3. Z(OCH_3), 4-methoxybenzyloxycarbonyl-; MTs, mesitylsulfonyl (cf. ref. 4) Met(O), methionine sulfoxide; Bzl, benzyl; z-, benzyloxycarbonyl; Tfa, trifluoroacetic acid.
4. Yajima H, Takeyama M, Kanaki J, Mitani K, J Chem Soc Chem Commun 1978: 482
5. Instead of the traditional but not always unequivocal Greek numbers (such as octacosapeptide in this case) the self-explanatory designation, 28-peptide is used (cf. Bodanszky M, Peptides Proc Fifth Amer Pept Symp, Goodman M, Meienhofer J eds, Wiley, New York 1977, p. 1).

6. Weak base-anion exchange resins such as Amberlite CG-4B or Dowex-1 (acetate) should be used.
7. In ref. 2 carboxymethylcellulose was applied and a gradient of ammonium hydrogen carbonate was used for elution. The peptide thus purified contained no significant impurities detectable by electrofocusing. The material from the Sephadex G-25 column was recovered in the electrofocusing experiment in 91% yield.
8. Deprotection in liquid hydrogen fluoride, in the presence of *m*-cresol, gave similar results.
9. For the acidolytic removal of protecting groups in the presence of thioanisole cf. also Kiso Y, Nakamura S, Ito K, Ukawa K, Kitagawa K, Akita T, Moritoki H, J Chem Soc Chem Commun 1979: 971; Kiso Y, Ukawa K, Nakamura S, Ito K, Akita T (1980) Chem Pharm Bull 28: 673
10. For the removal of the benzyloxycarbonyl group instead of the here described trifluoroacetic acid-thioanisole-*m*-cresol mixture, a 0.5 M solution of 4-(methyl-mercapto)phenol in trifluoroacetic acid can also be applied (Bodanszky M, Bodanszky A (1984) Int J Peptide Res 23: 287.

4.6 Liquid Hydrogen Fluoride [1, 2]

Oxytocin [1]

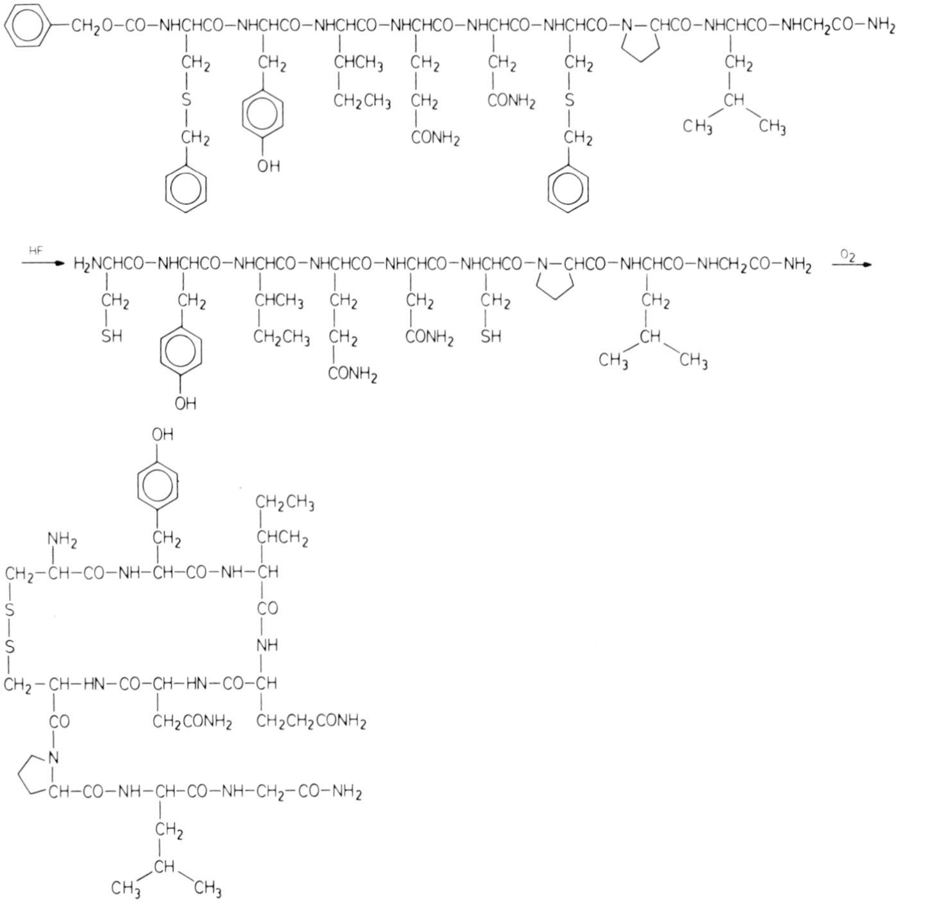

$C_6H_5-CH_2O-CO-NHCHCO-NHCHCO-NHCHCO-NHCHCO-NHCHCO-NHCHCO-N-CHCO-NHCHCO-NHCH_2CO-NH_2$

CH_2 CH_2 $CHCH_3$ CH_2 CH_2 CH_2 CH_2

S CH_2CH_3 CH_2 $CONH_2$ S CH

CH_2 $CONH_2$ CH_2 CH_3 CH_3

OH

$\xrightarrow{HF}$ $H_2NCHCO-NHCHCO-NHCHCO-NHCHCO-NHCHCO-NHCHCO-N-CHCO-NHCHCO-NHCH_2CO-NH_2$ $\xrightarrow{O_2}$

CH_2 CH_2 $CHCH_3$ CH_2 CH_2 CH_2 CH_2

SH CH_2CH_3 CH_2 $CONH_2$ SH CH

$CONH_2$ CH_3 CH_3

OH

OH

CH_2CH_3

NH_2 CH_2 $CHCH_2$

$CH_2-CH-CO-NH-CH-CO-NH-CH$

S CO

S NH

$CH_2-CH-HN-CO-CH-HN-CO-CH$

CO CH_2CONH_2 $CH_2CH_2CONH_2$

N

$CH-CO-NH-CH-CO-NH-CH_2-CO-NH_2$

CH_2

CH

CH_3 CH_3

oxytocin

A sample of the protected nonapeptide derivative *N*-benzyloxycarbonyl-*S*-benzyl-L-cysteinyl-L-tyrosyl-L-isoleucyl-L-glutaminyl-L-asparaginyl-*S*-benzyl-L-cysteinyl-L-prolyl-L-leucyl-glycinamide [3] (132 mg, 0.10 mmol) is dried at 80 °C over P_2O_5 for 3 hours in vacuo and then placed into an HF-resistant reaction vessel [4] together with anisole [5] (0.25 ml). Anhydrous hydrogen fluoride (about 5 ml) is distilled into the vessel and the resulting solution is kept at room temperature for 1 hour [6]. The hydrogen fluoride is removed by distillation under reduced pressure and the remaining material is kept in an evacuated desiccator over NaOH pellets for several hours. The deprotected peptide (hydrofluoride) is dissolved in distilled water (130 ml) [7], the pH adjusted to 6.5 with concentrated NH_4OH and CO_2 free air is passed through the solution [8] until the nitroprusside test becomes negative [9]. The pH is adjusted to 4 and the solution filtered from a small amount of insoluble material [10]: it is ready for bioassay [11].

1. Sakakibara S, Shimonishi Y (1965) Bull Chem Soc Jpn 38: 1412
2. Acidolysis of protecting groups with liquid HBr has been proposed (Brenner M, Curtius HC (1963) Helv Chim Acta 46: 2126). The solubility of proteins in liquid HF (Katz JJ (1954) Arch Biochem Biophys 51: 293) and the volatility of HF (b.p. 19.5 °C) prompted Sakakibara and Shimonishi (ref. 1) to explore its usefulness for the removal of protecting groups. A whole gamut of blocking groups is sensitive to HF and it also cleaves the bonds conventionally used for anchoring of peptides to insoluble polymers. Thus liquid HF is often applied for the cleavage of the already assembled peptide chain from the polymeric support (Lenard J, Robinson AB (1967) J Amer Chem Soc 89: 181).
3. du Vigneaud V, Ressler C, Swann JM, Roberts CW, Katsoyannis PG (1954) J Amer Chem Soc 76: 3115; Bodanszky M, du Vigneaud V (1959) J Amer Chem Soc 81: 5688
4. An apparatus constructed for this purpose (Sakakibara S, Shimonishi Y, Kishida Y, Okada M, Shugihara H (1964) Bull Chem Soc Jpn 40: 2164) is commercially available. The entire operation should be executed in a well ventilated hood.
5. Anisole may not be the ideal scavenger. While it is a good trap for carbocations it can participate in electrophilic aromatic substitution with the carboxyl groups in amino acid side chains or the C-terminal α-carboxyl group. It can also be the source of methyl groups which alkylate sensitive functions such as the thioether sulfur in methionine.
6. Removal of *S*-benzyl groups from sulfhydryl derivatives requires exposure to liquid HF at room temperature and a reaction time of about one hour. Under these conditions some $N \rightarrow O$ acyl migration can be expected in serine containing peptides. (Sakakibara S, Shin KH, Hess GP (1962) J Amer Chem Soc 84: 4921; Lenard J, Hess GP (1964) J Biol Chem 239: 3275). Most other protecting groups require less time and are removed at 0 °C and thus present less problems.
7. It is advisable to remove absorbed air from the water by prolonged boiling followed by cooling under nitrogen. The presence of air can cause premature oxidation: if disulfide formation takes place before the desired dilution is achieved, the formation of dimers and polymers will be observed.
8. Cyclization preferentially takes place at concentrations of 10^{-3} molar and less; at higher concentrations dimers and polymers are produced in significant amounts.
9. Sodium nitroprusside (sodium nitroferricyanide) gives a violet color with sulfhydryl derivatives. A grain of the solid reagent can be added to a drop of the solution. Solutions of the reagents are useful only for a short time. For a valid test the solution to be tested should be neutral or weakly alkaline.

10. At pH 4 oxytocin is fairly stable. Above pH 5 dismutation takes place with the formation of two dimers, oligomers and polymers. Polymers of oxytocin are insoluble in water, but the dimers remain in solution.
11. For details of the bioassay, isolation and purification of the hormone the papers cited in ref. 3 should be consulted.

5 Hydrolysis

5.1 Base Catalyzed Hydrolysis of Alkyl Esters (Saponification with Alkali) [1]

N-Benzyloxy-carbonyl-S-benzyl-L-cysteinyl-L-tyrosine [2]

$$C_6H_5-CH_2O-CO-NH-CH(CH_2-S-CH_2-C_6H_5)-CO-NH-CH(CH_2-C_6H_4-OH)-CO-OC_2H_5 \xrightarrow[2.\ HCl]{1.\ NaOH}$$

$$C_6H_5-CH_2O-CO-NH-CH(CH_2-S-CH_2-C_6H_5)-CO-NH-CH(CH_2-C_6H_4-OH)-COOH + C_2H_5OH$$

$C_{27}H_{28}N_2O_6S$ (508.6)

A solution of N-benzyloxycarbonyl-S-benzyl-L-cysteinyl-L-tyrosine ethyl ester [2, 3] (5.37 g, 10 mmol) in methanol (20 ml) is surrounded by a water bath of room temperature and N NaOH (22 ml) [4] is added with stirring. The mixture is stored at room temperature for two hours [5]. Dilute hydrochloric acid (10 ml, N HCl) is added and the methanol removed in vacuo. The aqueous solution is cooled in an ice-water bath and stirred during acidification to Congo (about 12 ml N HCl). After storage in the cold for two hours, the precipitate is collected on a filter, thoroughly washed with water and dried in air. Recrystallization from ethanol-water affords the dipeptide derivative (4.12 g, 81%) in analytically pure form, melting at 200–202 °C; $[\alpha]_D^{20} - 16°$ (c 4, pyridine).

1. Saponification of alkyl esters of peptides with alkali is a frequently used method for the removal of blocking groups from the C-terminal carboxyl group, but the procedure is far from unequivocal. Hydrolysis of methyl and ethyl esters can be accompanied by the loss of *tert*-butyl ester protection of aspartyl and glutamyl side chains, by the formation of dehydroalanine residues in serine-containing peptides or elimination of benzyl alcohol and hydantoin formation in benzyloxycarbonyl derivatives particularly in chains where glycine

occupies the second position in the sequence. Hydrolysis of carboxamides and several other side reactions have been encountered as was the loss of chiral integrity of the C-terminal residue which is directly involved in the process. These side reactions can be kept at a minimum by carrying out saponification at 0 °C rather than at room temperature, by adding the alkali in small portions and by avoiding excess alkali. Controlled hydrolysis at a predetermined pH (such as pH 10.5) accomplished with the aid of a pH stat or by monitoring alkalinity with thymolphthalein added to the mixture is also helpful. All these precautions notwithstanding saponification with alkali might be less than satisfactory with some peptide derivatives. In spite of such limitations methyl esters and ethyl esters are valuable protecting groups since they are readily converted to amides or hydrazides. If, however, the free carboxyl group is the aim of a process, it might be preferable to apply phenyl esters which are rapidly cleaved at pH 10.5 by the peroxide anion (H_2O_2 and NaOH) (Kenner GW, Seely JH (1972) J Amer Chem Soc 94: 3259).

2. Iselin B, Feurer M, Schwyzer R (1955) Helv Chim Acta 38: 1508; cf. also Harington CR, Pitt Rivers RV (1944) Biochem J 38: 417
3. The same protected dipeptide acid can be obtained through the saponification of the corresponding methyl ester as well (Bodanszky M, unpublished).
4. Only slightly more than the calculated amount of alkali is required for complete hydrolysis. In the present example one equivalent of NaOH is needed for the neutralization of the phenolic hydroxyl group.
5. Methyl esters are saponified more readily than ethyl esters. One hour reaction time (at room temperature) will be sufficient in most cases. Considerable variations in the rate of hydrolysis have been observed for different peptides. Thus, it is advisable to monitor the progress of the reaction, e.g. by recording the uptake of alkali (at constant pH) as a function of time.

5.2 Acid Catalyzed Hydrolysis [1]

Phthalyl-L-phenylalanyl-glycine [2]

Phthalyl-N-CH($CH_2C_6H_5$)-CO-NH-CH_2-CO-OC_2H_5 —HCl/HOH→ Phthalyl-N-CH($CH_2C_6H_5$)-CO-NH-CH_2-COOH + C_2H_5OH

$C_{19}H_{16}N_2O_5$ (352.3)

The protected dipeptide derivative phthalyl-L-phenylalanyl-glycine ethyl ester [2] (3.80 g, 10 mmol) is added to a mixture of acetone (40 ml), water (28 ml) and concentrated hydrochloric acid (12 ml) and the suspension is heated under reflux for 2 h. A clear solution results, which is evaporated to dryness under reduced pressure. The residue is dissolved in a solution of potassium hydrogen carbonate (4.8 g) in water (40 ml), the solution filtered and acidified to Congo with concentrated hydrochloric acid. Ethanol (20 ml) is added and the mixture heated to dissolve the precipitate. On cooling the product separates in fine colorless needles. The crystals are collected on a filter, thoroughly washed with water, dried in air and finally over P_2O_5 in vacuo. The protected dipeptide acid

(2.91 g, 83%) melts at 183–185 °C; $[\alpha]_D^{28}$ −148° (*c* 1.5, absolute ethanol) and is analytically pure.

1. The acid resistance of the phthalyl group permits the cleavage of alkyl esters by acid catalyzed hydrolysis but various protecting groups, the carboxamide groups in the side chains of asparagine and glutamine residues and the peptide bonds themselves are also hydrolyzed by hot aqueous acids. Thus, this method of deprotection is only suitable for certain combinations of protecting groups and for sequences which are not readily cleaved.
2. Sheehan JC, Chapman DW, Roth RW (1952) J Amer Chem Soc 74: 3822

5.3 Enzyme Catalyzed Hydrolysis

5.3.1 Enzymatic Removal of the Phenylacetyl Group from the ε-Amino Group of Lysine Residues [1]

Partial Deprotection of a Dipeptide Derivative

C_6H_5–CH_2–CO–NH–CH_2–CH_2–CH_2–CH_2–(–NH–CH–CO–) $\xrightarrow[\text{penicillin-amidohydrolase}]{\text{HOH}}$ –NH–CH(–$(CH_2)_4$–NH_2)–CO– + C_6H_5–CH_2COOH

A solution of N^α-benzyloxycarbonyl-N^ε-phenylacetyl-L-lysyl-glycinamide [2] (454 mg, 1 mmol) in acetic acid (5 ml) is treated with a 4 N solution of HBr in acetic acid (5 ml). After 30 min at room temperature the solution is diluted with ether (300 ml) and the solid hydrobromide salt is isolated by decantation. It is washed with ether (several times, by decantation) and dried in vacuo over NaOH pellets at room temperature. A sample of the salt (2.6 mg) is dissolved in a 0.1 M phosphate buffer of pH 7.5 (0.26 ml) and treated with a solution (0.135 ml) of penicillin-amidohydrolase [3] in the same buffer. The mixture is diluted with water (0.135 ml) and incubated at 37 °C. A sample taken after one and a half hours shows on thin layer chromatograms the complete disappearance of the starting material and the presence of a single ninhydrin positive compound with low R_f value.

Desamino Lysine Vasopressin

A solution of the cyclic peptide 1-desamino-8-N^ε-phenylacetyl-lysine vasopressin [1] (18 mg) in 96% ethanol (0.7 ml) is diluted with water (6.3 ml). A preparation of penicillin-aminohydrolase [3] (150 mg) is dissolved in a 0.1 M phosphate buffer of pH 7 (5 ml). A small amount of undissolved material is removed by centrifugation and the supernatant is added to the solution of the phenylacetyl-peptide. The reaction mixture is kept at 37 °C for one hour, then cooled to room temperature. Acidification to pH 3 with 0.01 M HCl coagulates the enzyme preparation and allows its removal by centrifugation. The solution is desalted on a column of Amberlite CG-50 (15 × 1 cm) and the product [4] is further purified by free-flow electrophoresis [5] to yield 12 mg of 1-desamino-8-lysine vasopressin in homogeneous form.

1. Bartnik F, Barth T, Jost K (1981) Collect Czechoslovak Chem Commun 46: 1983
2. For the preparation of this compound (by conventional methods) L-lysine was phenylacetylated [1] by the reaction if its Cu^{II}-complex with phenylacetyl chloride followed by decomposition of the copper complex with H_2S.
3. The enzyme preparation was obtained from *E. coli*. Cf. Kaufmann W, Bauer K (1960) Naturwissenschaften 47: 474; Rolinson GN, Batchelor FR, Butterworth D, Cameron-Wood J, Cole M, Eustace GE, Hart MV, Richards M, Chain EB (1960) Nature 187: 236; Claridge CA, Gourevitch A, Lein J (1960) Nature 187: 237. It seems to be likely that the enzyme will become commercially available.
4. This can be isolated by concentration of the solution in vacuo to a small volume and lyophylization.
5. Purification by ion exchange chromatography or by contercurrent distribution are probably equally suitable.

5.3.2 Chymotrypsin Catalyzed Hydrolysis of Alkyl Esters [1, 2]

L-Prolyl-L-phenylalanine [1]

$$\text{HN-CH-CO-NH-CH}(\text{CH}_2\text{C}_6\text{H}_5)\text{-CO-OCH}_3 \xrightarrow[\text{chymotrypsin}]{\text{HOH}} \text{HN-CH-CO-NH-CH}(\text{CH}_2\text{C}_6\text{H}_5)\text{-COOH} + \text{CH}_3\text{OH}$$

$C_{14}H_{18}N_2O_3$ (262.3)

To a stirred solution of L-prolyl-L-phenylalanine methyl ester (2.76 g, 10 mmol) in methanol (15 ml) a solution of α-chymotrypsin [3] (50 mg) in 0.5 M ammonium acetate (140 ml) adjusted to pH 6.4 is added rapidly, at room temperature. The product, L-prolyl-L-phenylalanine starts to crystallize immediately. Stirring is continued at room temperature for 15 minutes. The crystals are collected on a filter and washed with cold water (10 ml) and then with methanol (10 ml). The air dried dipeptide monohydrate [4] (2.5 g, 89%) is chromatographically homogeneous. Recrystallization from water and drying at 110 °C in vacuo yields an anhydrous product melting at 234 to 238 °C [5]; $[\alpha]_D^{22}$ −43° (*c* 2, 20% HCl).

Benzyloxy-carbonyl-L-valyl-L-tyrosine [1]

$$C_6H_5CH_2O\text{-}CO\text{-}NH\text{-}CH(CH(CH_3)_2)\text{-}CO\text{-}NH\text{-}CH(CH_2C_6H_4OH)\text{-}CO\text{-}OCH_3 \xrightarrow[\text{chymotrypsin}]{HOH}$$

$$C_6H_5CH_2O\text{-}CO\text{-}NH\text{-}CH(CH(CH_3)_2)\text{-}CO\text{-}NH\text{-}CH(CH_2C_6H_4OH)\text{-}COOH + CH_3OH$$

$C_{22}H_{26}N_2O_6 \cdot H_2O$ (432.5)

A solution of benzyloxycarbonyl-L-valyl-L-tyrosine methyl ester [5] (4.29 g, 10 mmol) in dimethylformamide (15 ml) is added dropwise, during a period of about 5 minutes, to a stirred solution of α-chymotrypsin [3] (170 mg) in 0.5 M ammonium acetate (200 ml), adjusted to pH 8 with ammonium hydroxide. Immediately a precipitate forms and a drop in the pH is observed. From time to time dilute ammonium hydroxide is added to readjust the pH to 8. After about two hours at room temperature a clear solution is obtained. Acidification with 6 N hydrochloric acid yields a crystalline precipitate which is collected on a filter, thoroughly washed with water and dried in air. Benzyloxycarbonyl-L-valyl-L-tyrosine monohydrate [6] (3.4 g, 79%) thus obtained is chromatographically homogeneous and melts at 163–169 °C; $[\alpha]_D^{22}$ +26.5 (*c* 2, pyridine) [7].

1. Walton E, Rodin JO, Stammer CH, Holly FW (1962) J Org Chem 27: 2255
2. Chymotrypsin acts as a catalyst specifically in the hydrolysis of peptide bonds between aromatic amino acid residues and the following residue (although, slow hydrolysis takes place at leucine residues as well). In the hydrolysis of ester bonds, however, specificity is greatly reduced and therefore the method is suitable for the hydrolysis of amino acid alkyl esters in general. Methyl, ethyl and even *tert*-butyl esters can be cleaved in this way, but only α-esters and at amino acid residues of L-configuration. The pH optimum might vary considerably from peptide to peptide. Similarly, trypsin can be used for the hydrolysis of esters and not only at basic residues. (Kloss G, Schröder E (1964) Hoppe Seyler's Z Physiol Chem 336: 248). A more recently proposed version of enzyme catalyzed hydrolysis of esters relies on the specific action of immobilized carboxypeptidase Y (Royer GP, Anantharamaiah GM (1979) J Amer Chem Soc 101: 3394).
3. Crystallized Worthington.
4. Fischer E, Luniak A (1909) Ber Dtsch Chem Ges 42: 4752
5. Schwarz H, Bumpus FM (1959) J Amer Chem Soc 81: 890
6. For the monohydrate a m.p. of 247 °C (corrected, 252 °C) was reported in ref. 4.
7. In ref. 6 a m.p. of 164–166 °C and the specific rotation $[\alpha]_D^{23}$ +23.3° (*c* 4, pyridine) are reported.

6 Base Catalyzed Elimination

6.1 Removal of the 9-Fluorenylmethyloxycarbonyl (Fmoc) Group [1]

9-Fluorenylmethyloxycarbonyl-L-asparaginyl-*S*-benzyl-L-cysteinyl-L-prolyl-L-tryprophyl-glycinamide [2]

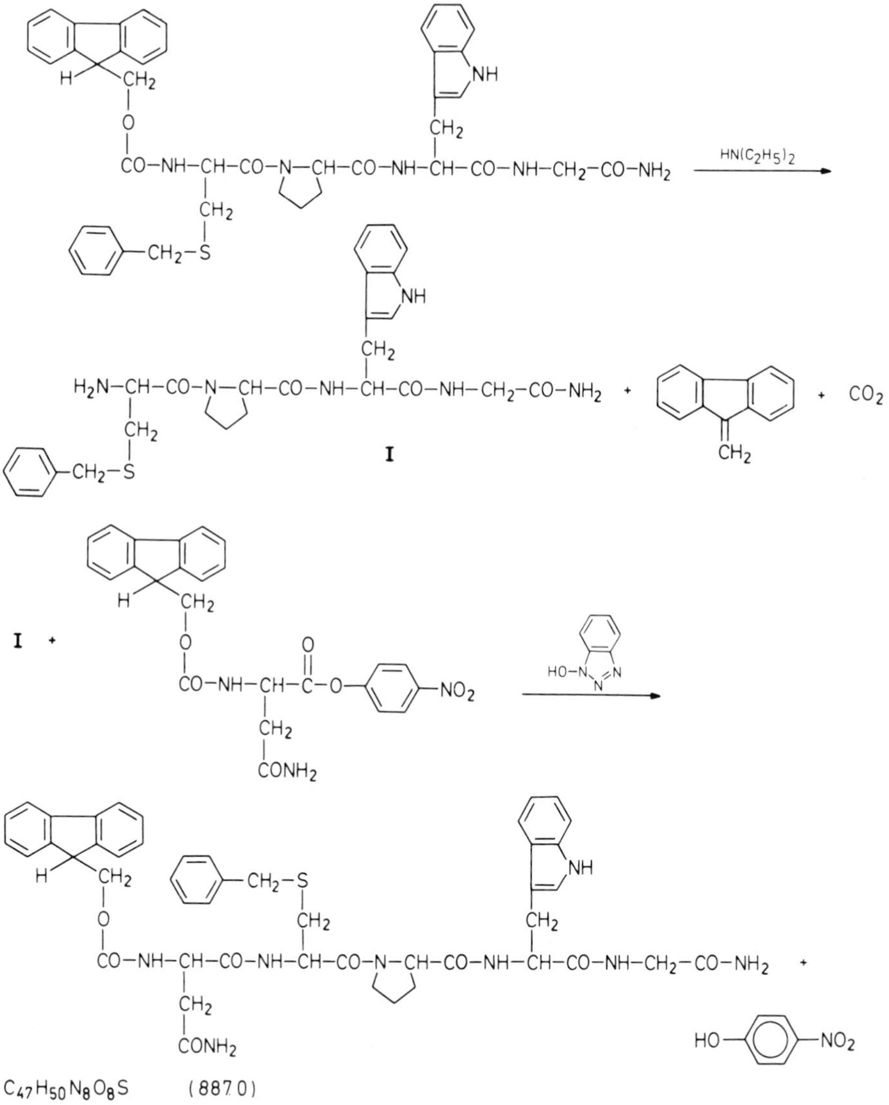

$HN(C_2H_5)_2$

$H_2N-CH-CO-N-CH-CO-NH-CH-CO-NH-CH_2-CO-NH_2$ + CO_2

I

I +

$CO-NH-CH-CO-NH-CH-CO-N-CH-CO-NH-CH-CO-NH-CH_2-CO-NH_2$ +

$C_{47}H_{50}N_8O_8S$ (887.0)

To a solution of Fmoc-Cys(Bzl)-Pro-Trp-Gly-NH_2 (773 mg, 1.0 mmol) in dimethylformamide (10 ml) diethylamine [3] (1.0 ml) is added and the reaction is allowed to proceed at room temperature for 1.5 hours. The diethylamine and the solvent are evaporated in vacuo at a bath temperature not exceeding 30 °C. The residue is triturated with ether (25 ml), the solid collected and washed with ether (twice, 20 ml each time) and dried in vacuo. The partially protected tetrapeptide amide is suspended [4] in a mixture of ethyl acetate (6 ml) and tetrahydrofurane (6 ml), 1-hydroxybenzotriazole (monohydrate, 153 mg, 1.0 mmol) is added followed by 9-fluorenylmethyloxycarbonyl-L-asparagine *p*-nitrophenyl ester [5] (666 mg, 1.4 mmol). The suspension is stirred at room temperature for two hours. During this time a change takes place in the appearance of the suspended solid and spot tests with ninhydrin and fluorescamine show the absence of the amino component. Ether (25 ml) is added, the product collected and washed with ether (three times, 20 ml each time) and dried in vacuo. The protected pentapeptide derivative (840 mg, 95%) melts at 183–184 °C dec.; $[\alpha]_D^{22} -47.5°$ (*c* 1, dimethylformamide); R_f 0.11 in $CHCl_3$–CH_3OH (9:1) and R_f 0.54 in EtOAc-pyridine-AcOH–H_2O (60:20:6:11). On elemental analysis correct values are found for C, H and N.

1. Carpino LA, Han GY (1970) J Am Chem Soc 92: 5478; (1972) J Org Chem 37: 3404 (1973) 38: 4318
2. Bodanszky M, Tolle JC, Bednarek MA, Schiller PW (1981) Int J Peptide Protein Res 17: 444
3. The more powerful piperidine (cf. ref. 1) was replaced by diethylamine, because it is more readily removed by evaporation. The elimination reaction is depicted here as leading to dibenzofulvene, carbon dioxide and a free amine (cf. ref. 1). It seems likely, however, that during evaporation of the solvent, the primarily generated diethylammonium carbamate is gradually converted to a carbamate comprising two moles of the amine: R–NH_3^+ · $^-$OOC–NH–R. This should have no unfavorable effect on the ensuing acylation reaction.
4. The commonly used solvent, dimethylformamide was not applied in this case in order to avoid the competing cyclization of Fmoc-Asn-ONp to the succinimide derivative.
5. Bodanszky A, Bodanszky M, Chandramouli N, Kwei JZ, Martinez J, Tolle JC (1980) J Org Chem 45: 72

6.2 Cleavage of 9-Fluorenylmethyl Esters [1]

L-Leucyl-L-phenylalanine [2]

$HN(C_2H_5)_2$

$H_2N-CH(CH_2CH(CH_3)_2)-CO-NH-CH(CH_2C_6H_5)-COOH + CO_2 + 2$ dibenzofulvene

$C_{15}H_{22}N_2O_3$ (278.3)

The fully protected dipeptide (0.68 g 1.0 mmol) is dissolved in a mixture of dimethylformamide (9 ml) and diethylamine (1.0 ml) and the solution is allowed to stand at room temperature for two hours. The amine and the solvent are removed in vacuo at a bath temperature not exceeding 30 °C. The residue is triturated with a mixture of ether (3 ml) and hexane (12 ml), the solid product collected on a filter and washed with a mixture of ether (5 ml) and hexane (5 ml). The free dipeptide (0.26 g, 94%) starts to decompose at 235° and melts with further decomposition at 252–255 °C; $[\alpha]_D^{23} + 37°$ (*c* 1.4, acetic acid). It gives a single ninhydrin positive spot on thin layer chromatograms (silica gel) with R_f 0.49 in the solvent system *n*-butanol-acetic acid-water (4:1:1).

1. Bednarek MA, Bodanszky M (1983) Int J Pept Protein Res 21: 196
2. In this example the amine protecting Fmoc group and the 9-fluorenylmethyl (Fm) group used for the blocking of the carboxyl function are simultaneously removed by the action of diethylamine via proton abstraction from C_9 of the fluorene system followed by elimination of dibenzofulvene.

7 Nucleophilic Displacement

7.1 Removal of the *o*-Nitrobenzenesulfonyl (Nps) Group [1, 2, 3]

***N*^ε^-*tert*-Butyloxycarbonyl-L-lysyl-L-leucyl-L-phenylalanyl-*N*^ε^-*tert*-butyloxycarbonyl-L-lysyl-*N*^ε^-*tert*-butyloxycarbonyl-L-lysine**

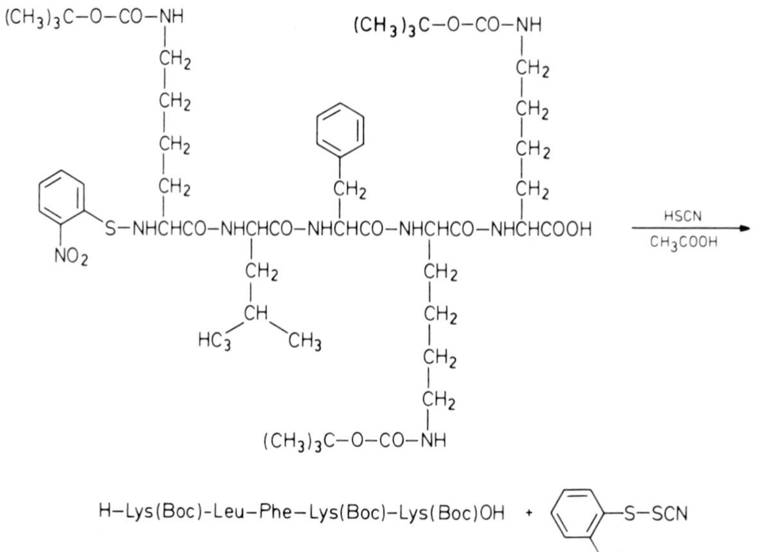

$\xrightarrow[CH_3COOH]{HSCN}$

H–Lys(Boc)–Leu–Phe–Lys(Boc)–Lys(Boc)OH + (o-NO₂-C₆H₄)–S–SCN

$C_{48}H_{82}N_8O_{12}$ (963.2)

(o-NO₂-C₆H₄)–S–SCN + 2-methylindole ⟶ 3-(o-nitrophenylthio)-2-methylindole + HSCN

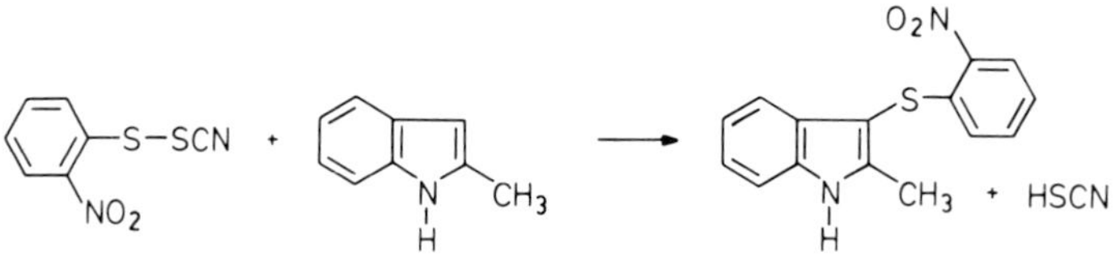

To a solution of ammonium rhodanide (0.16 g, 2 mmol) and 2-methyl-indole [4] (0.26 g, 2.1 mmol) in methanol (6 ml) and acetic acid (18 ml) Nps-Lys(Boc)-Leu-Phe-Lys(Boc)-Lys(Boc)OH (1.12 g, 1.0 mmol) is added and the reaction mixture is allowed to stand at room temperature for 3 hours. The solvent is removed under reduced pressure, the residue washed with warm (about 50 °C) distilled water (twice, 10 ml each time), with 2 N ammonium hydroxide (twice, 10 ml each time), triturated with ether and transferred to a filter. Trituration and washing with ether are continued until a white product is obtained. The partially deprotected pentapeptide is dried in vacuo over P_2O_5 and KOH pellets. It weighs 0.87 g (90%) and melts, with decomposition, at 217 °C; $[\alpha]_D^{20}$ −4.2° (*c* 1.2, acetic acid). On elemental analysis satisfactory values are obtained for C, H and N.

1. Wünsch E, Spangenberg R (1972) Chem Ber 105: 740
2. From the numerous nucleophiles proposed for the displacement of the Nps group rhodanide was selected for this volume because the reaction was efficiently executed in a simple procedure (ref. 1). The presence of an acceptor such as 2-methylindole, is essential for complete deprotection.
3. Instead of the commonly used expression *o*-nitrophenylsulfenyl the more appropriate term *o*-nitrobenzenesulfenyl is used here.
4. Because of the unpleasant odor of 2-methylindole the reaction should be carried out in a well ventilated hood and the filtrates should be disposed with proper care. A less volatile indole derivative such as acetyl-L-tryptophan *n*-butyl ester (Bodanszky M, Tolle JC, Seto J, Sawyer WH (1980) J Med Chem 23: 1258 might be equally suitable.

7.2 Cleavage of 2-Trimethylsilylethyl Esters with Fluorides [1, 2]

N-Benzyloxy-carbonyl-O-tert-butyl-L-threonine

$$C_6H_5-CH_2O-CO-NH-CH(CH(CH_3)-OC(CH_3)_3)-CO-O-CH_2-CH_2-Si(CH_3)_3 + (C_4H_9)_4\overset{+}{N}\cdot\overset{-}{F} \longrightarrow$$

$$C_6H_5-CH_2O-CO-NH-CH(CH(CH_3)-OC(CH_3)_3)-COO^{-}\,{}^{+}N(C_4H_9)_4 \quad (CH_2{=}CH_2 + (CH_3)_3SiF) \xrightarrow{HCl}$$

$$C_6H_5-CH_2O-CO-NH-CH(CH(CH_3)-OC(CH_3)_3)-COOH + (C_4H_9)_4\overset{+}{N}\cdot\overset{-}{Cl}$$

$C_{16}H_{23}NO_5$ (309.4)

A solution of *N*-benzyloxycarbonyl-*O*-*tert*-butyl-L-threonine 2-trimethylsilylethyl ester [3] (0.41 g, 1 mmol) in dimethylformamide (9 ml) is stirred at room temperature and treated with a freshly prepared 2.1 molar solution of tetrabutylammonium fluoride [4] in dimethylsulfoxide (1.2 ml, 2.5 mmol). A vigorous evolution of gas can be observed. Three minutes later the mixture is cooled in an ice-water bath, diluted with distilled water (10 ml) and concentrated to a small volume under reduced pressure. The remaining solution is diluted with ethyl acetate (50 ml) and enough 2 N HCl is added to render the lower layer acidic to Congo. The ethyl acetate layer is washed several times in water until a drop of the last wash leaves no residue on evaporation. The product is secured by evaporation of the solvent in vacuo. The yield is quantitative (0.31 g) [5].

1. Sieber P (1977) Helv Chim Acta 60: 2711
2. The method of cleavage of trimethylsilylethyl esters is based on a procedure developed for the removal of the 2-trimethylsilylethyloxycarbonyl amine protecting group (Carpino LA, Tsao JH, Ringsdorf H, Fell E, Hettrich G (1978) J Chem Soc Chem Commun 358).
3. 2-Trimethylsilylethyl esters of protected amino acids are readily obtained. To a solution of the acid (10 mmol) in acetonitrile (about 8 ml) pyridine (1.6 ml) and 2-trimethylsilylethanol (1.7 ml) are added and the stirred mixture is cooled in an ice-water bath. Ten minutes later dicyclohexylcarbodiimide (2.25 g) is added and stirring continued at 0 °C for an additional hour. The reaction mixture is stored in a refrigerator overnight, treated with a 5 M solution of oxalic acid in dimethylformamide (0.3 ml), and 30 min later filtered from the precipitate and the latter is washed with ethyl acetate. The filtrate is extracted with 0.1 N HCl, with 0.5 N $KHCO_3$, the solvent evaporated in vacuo and the residue chromatographed on a silica gel column with mixtures of hexane and ethyl acetate as eluents. The majority of esters are oils.
   Specific rotations of several esters are described in ref. 1. – If a protected amino acid is not sufficiently soluble in acetonitrile a small amount of dimethylformamide is added to achieve its dissolution.
4. Commercially available.
5. On thin layer chromatograms (silicagel), in the system chloroform-acetic acid (98:2) a single spot is detected, with R_f 0.25. No allothreonine derivative (R_f 0.19) is present. Thus the process does not lead to racemization.

$$2(CH_3)_3C{-}O{-}CO{-}NH{-}CH(CH_2{-}S{-}C(C_6H_5)_3){-}CO{-}NH{-}CH_2{-}CO{-}NH{-}CH(CH_2{-}CH_2{-}CO{-}O{-}C(CH_3)_3){-}CO{-}O{-}C(CH_3)_3 + 2I_2 + 2CH_3OH \longrightarrow$$

$$[(CH_3)_3C{-}O{-}CO{-}NH{-}CH(CH_2{-}S{-}){-}CO{-}NH{-}CH_2{-}CO{-}NH{-}CH(CH_2{-}CH_2{-}CO{-}O{-}C(CH_3)_3){-}CO{-}O{-}C(CH_3)_3]_2 + 2(C_6H_5)_3C{-}O{-}CH_3 + 2HI$$

$C_{46}H_{80}N_6O_{16}S_2$ (1037.3)

Removal of the *S*-Trityl Group by Iodolysis with Concomitant Oxidation to the Disulfide [1, 2]

N-tert-Butyloxycarbonyl-*S*-trityl-L-cysteinyl-glycyl-L-glutamic acid di-*tert*-butyl ester (1.53 g, 2 mmol) and iodine (508 mg, 2 mmol) are dissolved in methanol (25 ml) and allowed to react at room temperature for one hour. The mixture is cooled in an ice-water bath and decolorized by the dropwise addition of a one molar solution of sodium thiosulfate in water. The mixture is diluted with water (50 ml) and the precipitated material filtered and dried. It is extracted with petroleum ether, three times (10 ml each time) [3]. The remaining solid is crystallized from ethyl acetate–hexane: 945 mg (91%); it melts at 150–152 °C. On elemental analysis correct values are obtained for C, H, N and S [4].

1. Kamber B, Rittel W (1968) Helv Chim Acta 51: 2061. The original publication describes the preparation of mixed disulfides as well.
2. If a free thiol is the desired product, cleavage with acids in the presence of a large excess of mercaptane can be used (cf. König W, Kernbeck K, J Liebigs Ann Chem 1979: 227). The *S*-trityl peptide (10 mmol) is dissolved in a well stirred mixture of ethanethiol (25 ml) and

trifluoroacetic acid (25 ml). Removal of the *S*-trityl group is complete within 30 min at room temperature. If in the same operation also *tert*-butyl groups are to be cleaved, the reaction time should be extended to four hours. With high molecular weight substrates, the indicated amounts of ethanethiol and trifluoroacetic acid must be increased in order to dissolve the protected intermediate.

3. The petroleum ether extracts contain almost the calculated amount of methyl triphenylmethyl ether.
4. Under the conditions used in this cleavage the *S*-benzyl group remains intact.

9 Reduction of Methionine Sulfoxide Containing Peptides [1]

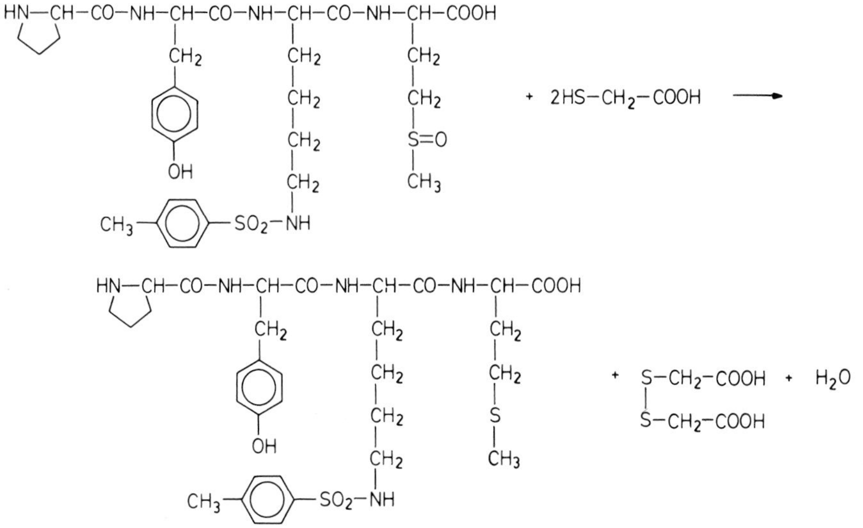

L-Prolyl-L-tyrosyl-N^{ε}-tosyl-L-lysyl-L-methionine [1]

HN–CH–CO–NH–CH–CO–NH–CH–CO–NH–CH–COOH + 2HS–CH$_2$–COOH ⟶

HN–CH–CO–NH–CH–CO–NH–CH–CO–NH–CH–COOH + S–CH$_2$–COOH / S–CH$_2$–COOH + H$_2$O

$C_{32}H_{45}N_5O_8S_2$ (691.9)

Freshly distilled thioglycolic acid [2, 3] (0.92 g, 0.70 ml, 10 mmol) is added to a suspension of L-prolyl-L-tyrosyl-N^{ε}-tosyl-L-lysyl-L-methionine-sulfoxide (0.71 g, 1.0 mmol) in distilled water (5 ml) and the reaction mixture is stored under nitrogen at 50 °C. A clear solution forms from which the product gradually separates as an oil. After about 20 h the mixture is evaporated in vacuo and the residual oil triturated with ether, the ether decanted and replaced with fresh ether. Trituration is continued until a solid material is obtained. This is dissolved in a small volume of methanol and the solution made slightly alkaline by the addition of a 2.5 M solution of NH_3 in methanol. The mixture is cooled in an ice water bath and the crystalline product which separates from the solution is collected after about 30 min. It is extracted with hot methanol and refiltered. The reduced (still partially protected) tetrapeptide (0.60 g, 87%) melts at 223–226 °C; $[\alpha]_D^{26} - 14°$ (c 4, acetic acid). It is chromatographically homogeneous and gives the expected values on elemental analysis.

1. Iselin B (1961) Helv Chim Acta 44: 61
2. Commercially available as mercaptoacetic acid.

3. Several thiols were recommended for the reduction of the sulfoxide group, such as thiophenol (Yajima H, Funakoshi S, Fuji N, Akaji K, Irie H (1979) Chem Pharm Bull 27: 1060) and dithiothreitol (Polzhofer KP, Ney KH (1971) Tetrahedron 27: 1997). A detailed study of the problem (Houghten RA, Li CH (1979) Anal Biochem 98: 36) led to the selection of *N*-methylmercaptoacetamide $HS–CH_2–CO–NHCH_3$, as a particularly suitable reducing agent.

V Special Procedures

1 Ammonolysis of Esters

1.1 Conversion of a Peptide Ethyl Ester to the Corresponding Amide

Benzyloxy-carbonyl-L-prolyl-L-leucyl-glycinamide [1]

$$C_6H_5-CH_2-O-CO-N(\text{pyrrolidine})-CH-CO-NH-CH(CH_2-CH(CH_3)_2)-CO-NH-CH_2-CO-OC_2H_5 \xrightarrow{NH_3}$$

$$C_6H_5-CH_2-O-CO-N(\text{pyrrolidine})-CH-CO-NH-CH(CH_2-CH(CH_3)_2)-CO-NH-CH_2-CO-NH_2 + C_2H_5OH$$

$C_{21}H_{30}N_4O_5$ (418.5)

The protected tripeptide derivative benzyloxycarbonyl-L-prolyl-L-leucyl-glycine ethyl ester [1] (44.8 g, 100 mmol) is dissolved in methanol (750 ml) with gentle warming. A vigorous stream of NH_3 gas is introduced [2] while the reaction mixture is cooled in an ice-water bath. After about one and a half hours introduction of the ammonia is discontinued and the flask is closed with a glass stopper. The solution is allowed to warm up to room temperature [3] and left to stand in a hood overnight [4]. The solvent is removed in vacuo or more conveniently by spontaneous evaporation from a large crystallizing dish placed in a hood [5]. The dry residue is triturated with ethyl acetate (200 ml), collected on a filter, washed with ethyl acetate (300 ml) and dried in air. The protected tripeptide amide (38 g, 90%) shrinks at 160 °C and melts at 162–164 °C. Recrystallization from a large volume of ethyl acetate leaves the melting point unchanged.

1. Ressler C and du Vigneaud V (1954) J Amer Chem Soc 76: 3107
2. A round bottom flask provided with a gas-inlet-outlet tube is best suited for the reaction. The entire operation should be carried out in a well ventilated hood.
3. Some pressure develops and it is advisable to wrap the flask in a heavy cloth. Also, the stopper should be secured, but not too tightly. A moderately heavy rubber stopper or other weight will prevent the inadvertent removal of the glass stopper.

4. Samples taken from the mixture from time to time show that conversion of the ester to the amide at room temperature requires only a few hours. If the reaction is allowed to proceed for several days a product with lower melting point is obtained.
5. Evaporation can be accelerated by a stream of air.

1.2 Ammonolysis of Active Esters

Benzyloxy-carbonyl-L-asparagine Amide [1]

$$C_6H_5{-}CH_2{-}O{-}CO{-}NH{-}CH(CH_2{-}CO{-}NH_2){-}\overset{O}{\overset{\|}{C}}{-}O{-}C_6H_4{-}NO_2 \xrightarrow{NH_3} C_6H_5{-}CH_2{-}O{-}CO{-}NH{-}CH(CH_2{-}CO{-}NH_2){-}CO{-}NH_2 + HO{-}C_6H_4{-}NO_2$$

$C_{12}H_{15}N_3O_4$ (265.3)

Dry ammonia is led [2] in a gentle stream over the stirred solution of benzyloxycarbonyl-L-asparagine *p*-nitrophenyl ester [3] (3.87 g, 10 mmol) in tetrahydrofurane (100 ml). The solution turns yellow and after a few minutes a white solid starts to separate. Two hours later the crystals are collected on a filter and washed with tetrahydrofurane [4] (75 ml) and dried in air. The amide (2.60 g, 98%) melts at 225–226 °C dec. [5].

1. Bodanszky M, Klausner YS and Mutt V (1972) Bioorg Chem 2: 30
2. The operations should be carried out in a well ventilated hood.
3. Bodanszky M, Denning GS, Jr, and du Vigneaud V (1963) Biochem Prep 10: 122
4. If the product has orange color from coprecipitated ammonium *p*-nitrophenolate the latter should be decomposed by the addition of a small amount of acetic acid to the tetrahydrofurane and ethyl acetate used for washing.
5. Sondheimer E and Holley RW (1954) J Amer Chem Soc 76: 2467 reported 219–223 °C.

2 Transesterification

2.1 Cleavage of the Ester Bond Between Peptide and Polymeric Support by Transesterification Followed by Hydrolysis [1]

Benzyloxycarbonyl-L-alanyl-L-phenylalanine

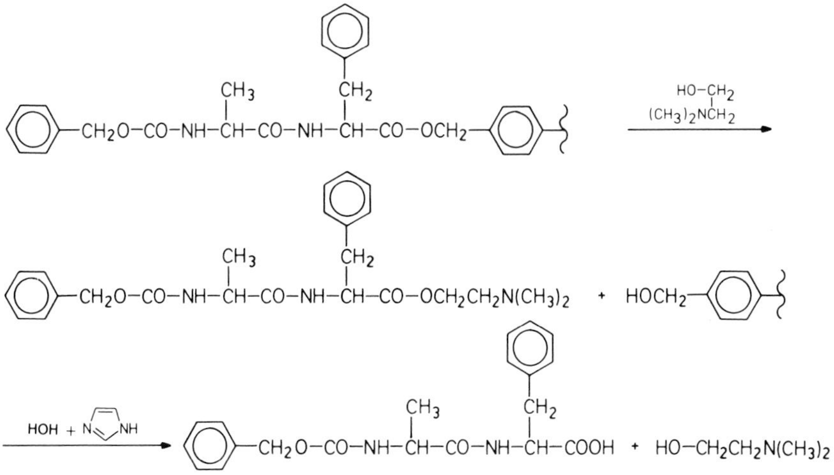

$C_{20}H_{22}N_2O_5$ (370.4)

The peptidyl polymer (*Z*-Ala-Phe-resin, 4.0 g, 1 mmol) is washed with dimethylformamide [2] (40 ml, in several portions) and then suspended, without drying in a mixture of dimethylformamide (80 ml) and dimethylaminoethanol (80 ml). The suspension is stirred for about 24 hours, filtered and the resin washed with dimethylformamide (three times, 80 ml each time). The filtrate and washings are pooled and concentrated in vacuo at room temperature to about 10 ml. An equal volume of water and a catalytic amount of imidazole (about 7 mg, 0.1 mmol) are added and the mixture allowed to stand at room temperature for one day [3]. The solvents are removed in vacuo at room temperature and the residue dissolved in ethyl acetate (100 ml). The solution is washed with 0.01 N hydrochloric acid (twice, 50 ml each time) and the aqueous phases reextracted with ethyl acetate (twice, with 50 ml each time). The ethyl acetate solutions are pooled, dried over anhydrous $MgSO_4$ and evaporated to dryness in vacuo. The crude product is recrystallized from aqueous acetic acid.

Benzyloxycarbonyl-L-alanyl-L-phenylalanine (0.28 g; 75%) is obtained with a m.p. of 125 °C (softening at 115 °C). Recrystallization from ethyl acetate-hexane raises the m.p. to 124–126 °C.

1. Barton MA, Lemieux RU, Savoie JY (1973) J Amer Chem Soc 95: 4501
2. The washings are discarded.
3. The hydrolysis is monitored by thin layer chromatograms.

3 Cyclization

3.1 Cyclization Through the Formation of a Disulfide Bond

Desamino-oxytocin [1]

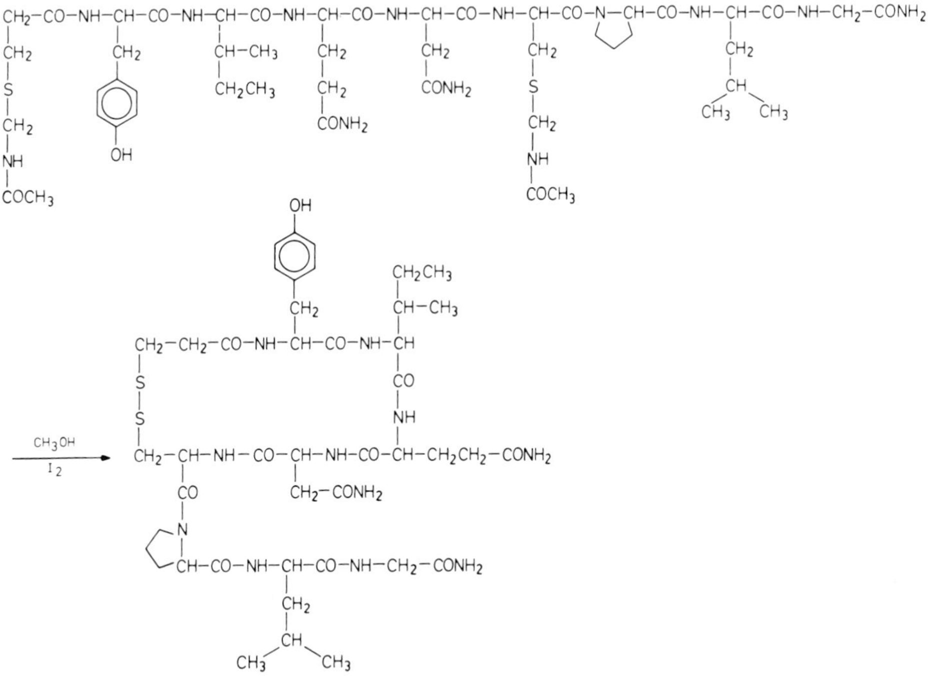

CH_2–CO–NH–CH–CO–NH–CH–CO–NH–CH–CO–NH–CH–CO–NH–CH–CO–N–CH–CO–NH–CH–CO–NH–CH_2–$CONH_2$

CH_2 | CH_2 | CH–CH_3 | CH_2 | CH_2 | CH_2 | CH_2

S | (C₆H₄) | CH_2CH_3 | CH_2 | $CONH_2$ | S | CH

CH_2 | OH | $CONH_2$ | CH_2 | CH_3 CH_3

NH | NH

$COCH_3$ | $COCH_3$

$\xrightarrow[I_2]{CH_3OH}$

OH

CH_2CH_3

CH_2 CH–CH_3

CH_2–CH_2–CO–NH–CH–CO–NH–CH

S CO

S NH

CH_2–CH–NH–CO–CH–NH–CO–CH–CH_2CH_2–$CONH_2$

CO CH_2–$CONH_2$

N

CH–CO–NH–CH–CO–NH–CH_2–$CONH_2$

CH_2

CH

CH_3 CH_3

$C_{43}H_{65}N_{11}O_{12}S_2 \cdot 3H_2O$ (1046.6)

The protected peptide amide, 1-desamino-*S*,*S*′-diacetamidomethyl-oxytocein [1] (115 mg, 0.10 mmol) is dissolved in 75% aqueous methanol (600 ml) and a solution of iodine (64 mg, 0.25 mmol) in methanol (150 ml) is added, with stirring, during one hour [2]. After four more hours at room temperature the solution is concentrated in vacuo at a bath temperature of 30 °C to about 25–30 ml. The concentrated solution is filtered through a Sephadex G10 column: elution of the peptide is monitored by u.v. absorption at 275 nm. The peptide is further purified by countercurrent distribution in the solvent system *n*-butanol-toluene –0.05% acetic acid (3:2:5) [3]. The target compound migrates with a K value of 0.5. It is recovered by concentration of the solution in vacuo and liophilization of the concentrated solution: 55 mg. Crystallization from water (2 ml) in the presence of a trace of acetic acid affords the trihydrate (34 mg; 32.5%) melting at 174–175 °C; $[\alpha]_D -90°$ (*c* 0.5, 1 M AcOH). It is homogeneous in chromatography on thin layer plates of silica gel, R_f 0.35 (*n*-butanol-acetic acid-water, 4:1:1); R_f 0.43 (chloroform-methanol, 1:1). Correct values are obtained on elemental analysis and the expected ratios on amino acid analysis. The synthetic peptide has 600 IU uterotonic activity per mg.

1. Marbach P, Rudinger J (1974) Helv. Chim. Acta 57: 403
2. Kamber B (1971) Helv Chim Acta 54: 927
3. Hope D, Murti VVS, du Vigneaud V (1962) J Biol Chem 237: 1563

3.2 Ring Closure Through an Amide Bond

Ditosyl-gramicidin S [1]

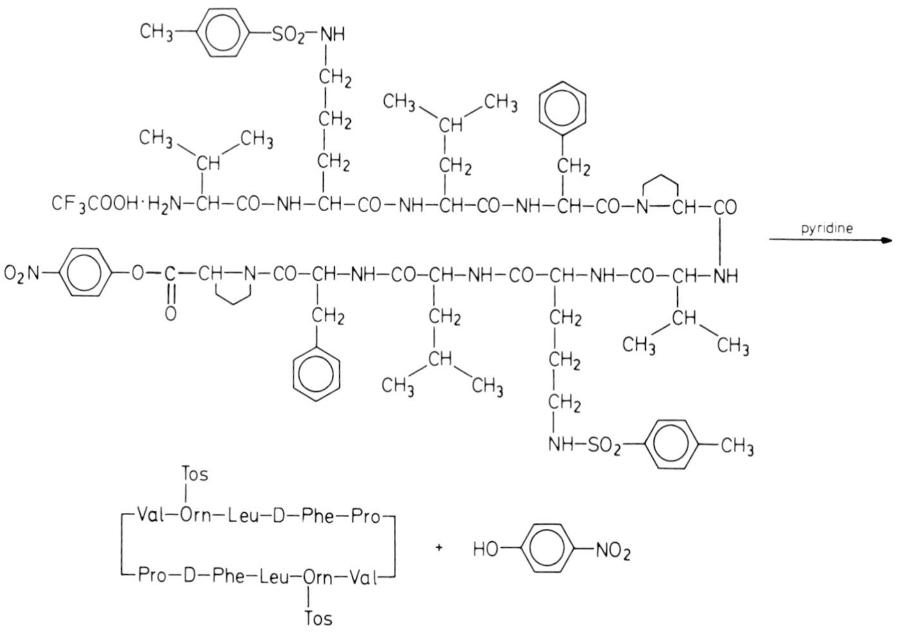

$CF_3COOH \cdot H$-Val-Orn(Tos)-Leu-D-Phe-Pro-Val-Orn(Tos)-Leu-D-Phe-Pro-ONp $\xrightarrow{\text{pyridine}}$

cyclo(-Val-Orn(Tos)-Leu-D-Phe-Pro-Val-Orn(Tos)-Leu-D-Phe-Pro-) + HO–C₆H₄–NO₂

$C_{74}H_{104}N_{12}O_{14}S_2 \cdot 2H_2O$ (1485.8)

A solution of the trifluoroacetate salt of *H*-Val-Orn(Tos)-Leu-D-Phe-Pro-Val-Orn(Tos)-Leu-D-Phe-Pro-ONp (0.39 g, 0.20 mmol) [2] in dimethylformamide (10 ml) and acetic acid (3 drops) is added dropwise to pyridine (75 ml), which is stirred and kept at 55 °C. The addition of the active ester salt should take about 4 h; stirring and heating to 55 °C are continued for an additional hour [3]. The solvent is removed under reduced pressure, the residue extracted, under reflux, with a boiling mixture of ether (25 ml) and hexane (25 ml) and then similarly with boiling ether (50 ml). The extracted powder is dissolved in a mixture of isopropanol, methanol and water (10 ml each) and filtered through a column of a strongly basic ion-exchange resin and then through a strongly acidic ion-exchanger [4]. The filtrate is warmed to 45 °C and diluted with water until no more product separates. The precipitate is collected on a filter and dried in vacuo. This crude material (170 mg) is purified by chromatography on neutral alumina. The first eluates, obtained with benzene-chloroform (9:1) are discarded. From the fractions eluted with mixtures of chloroform and ethyl acetate an almost colorless product (145 mg) is secured which is crystallized from 65% ethanol. Di-tosyl-gramicidin S, *cyclo* [Val-Orn(Tos)-Leu-D-Phe-Pro-Val-Orn(Tos)-Leu-D-Phe-Pro-], (92 mg, 31%) [5] forms long prisms which melt with decomposition at 318 °C [6]; $[\alpha]_D^{24} -188°$ (*c* 0.7, acetic acid). A sample dried at 110 °C and 10^{-2} mm for 2 h gives C, H and N values that correspond to the dihydrate. After overnight drying under the same conditions the analytical values agree with those calculated for the monohydrate.

1. Schwyzer R, Sieber P (1957) Helv Chim Acta 40: 624
2. This crude *p*-nitrophenyl ester (salt) was obtained through the esterification of the corresponding N^{α}-trityl decapeptide with the help of di-*p*-nitrophenyl sulfite followed by the removal of the trityl group with trifluoroacetic acid. The ester-salt was about 90% pure.
3. Completeness of the reaction can be ascertained by the evaporation of a small sample in vacuo and trituration of the residue with ether. The ether extract turns yellow when treated with N NaOH; the product should not give a yellow color.
4. No details are given in the original about the purification by ion-exchange.
5. Calculation of the yield is based on the assumption that the active ester (salt) is 90% pure.
6. Countercurrent distribution in the system tetrachloromethane -85% methanol (1:1) affords a product which darkens at 305 °C and melts with decomposition at 319 °C. The K value of ditosyl-gramicidin S is 0.35 in this system.

Desthiomalformin [1]

$C_6H_5CH_2O-CO-NH-CH(CH(CH_3)CH_2CH_3)-CO-NH-CH(CH_3)-CO-NH-CH(CH_3)-CO-NH-CH(CH(CH_3)_2)-CO-NH-CH(CH_2CH(CH_3)_2)-CO-NH-NH_2$

$\xrightarrow{HBr/AcOH}$ $HBr \cdot H_2N-CH(CH(CH_3)CH_2CH_3)-CO-NH-CH(CH_3)-CO-NH-CH(CH_3)-CO-NH-CH(CH(CH_3)_2)-CO-NH-CH(CH_2CH(CH_3)_2)-CO-NHNH_2$

$\xrightarrow{HONO}$ $HBr \cdot H_2N-CH(CH(CH_3)CH_2CH_3)-CO-NH-CH(CH_3)-CO-NH-CH(CH_3)-CO-NH-CH(CH(CH_3)_2)-CO-NH-CH(CH_2CH(CH_3)_2)-CO-N=\overset{+}{N}=\overset{-}{N}$

$\xrightarrow{CH_3CH_2-N(CH(CH_3)_2)-CH(CH_3)_2}$ cyclo (D-Leu → L-Ile – D-Ala – D-Ala – L-Val –)

D-Leu

L-Ile L-Val

D-Ala–D-Ala

$C_{23}H_{41}N_5O_5$ (467.6)

To a suspension of benzyloxycarbonyl-L-isoleucyl-D-alanyl-D-alanyl-L-valyl-D-leucine hydrazide [1] (0.64 g, 1.0 mmol) in acetic acid (5 ml) in a 250 ml round bottom flask an about 4.5 M solution of HBr in acetic acid (5 ml) is added. The flask is closed with a cotton filled drying tube and swirled to dissolve the hydrazide. The solution is allowed to stand at room temperature for about one hour. Dilution with ether (150 ml) precipitates the hyrdrobromide salt of the pentapeptide hydrazide. It is collected on a sinter-glass filter thoroughly washed with ether (100 ml) and dried over NaOH pellets in vacuo [2]. The dry material is dissolved in dimethylformamide (10 ml), the solution cooled in a dry ice-acetone bath to −20 °C, treated with conc. HCl (0.10 ml) followed by a 1 M solution of $NaNO_2$ in water (1.0 ml), added dropwise with swirling. After about 15 minutes at about −15 °C the reaction mixture is diluted [3] with dimethylformamide (200 ml) precooled to the same temperature, treated with diisopropylethylamine (1.4 ml) [4], and the solution is stored at 4 °C. Already during the first hours a white crystalline solid separates. After about a day the crystals are collected on a filter, washed with dimethylformamide (20 ml) and with ethyl acetate (20 ml) and dried in vacuo. The product, *cyclo*(L-isoleucyl-D-alanyl-D-alanyl-L-valyl-D-leucyl-), weighs 0.40 g (86%) [5] does not melt below 300 °C; $[\alpha]_D^{25}$ +55.6° (*c* 1, trifluoroacetic acid). It appears as a single opaque spot on thin layer chromatograms sprayed with water; chloroform-methanol (9:1), *n*-butanol-acetic acid-water (4:1:1) or neat trifluoroethanol can be used as eluents. Both on amino acid analysis and on elemental analysis the calculated values are obtained. The cyclopentapeptide sublimes unchanged at 260–290 °C at ca 10 Pa.

1. Bodanszky M, Henes JB (1975) Bioorg Chem 4: 212
2. The precipitate is not the monohydrobromide: it contains several moles of HBr which are gradually lost on drying in vacuo over NaOH. Removal of the excess HBr is, however, not necessary for the next step.
3. In concentrated solutions dimerization and polymerization compete with cyclization and cyclizations are best carried out in dilute solutions (e.g. 10^{-3} M) to favor the unimolecular reaction. In the case of desthiomalformin the alternation of D and L residues leads to a quasicyclic preferred conformation in the open chain precursor; hence, very little cyclodimerization takes place and the reaction can be carried out with almost the same results if dilution with dimethylformamide is omitted.
4. A large excess of tertiary amine should be avoided. The base is added in small portions until a moist indicator paper held close to the surface of the solution shows an alkaline reaction.
5. In most cyclization reactions lower yields were observed. The high yield in the ring closure described in this example is probably due to a favorable conformation of the open chain precursor. Also, the azide method seems to be well suited for cyclization.

4 Polycondensation

4.1 *N*-Carboxyanhydrides [1]

N^{ε}-Benzyloxy-carbonyl-N^{α}-carboxy-L-lysine Anhydride [2]

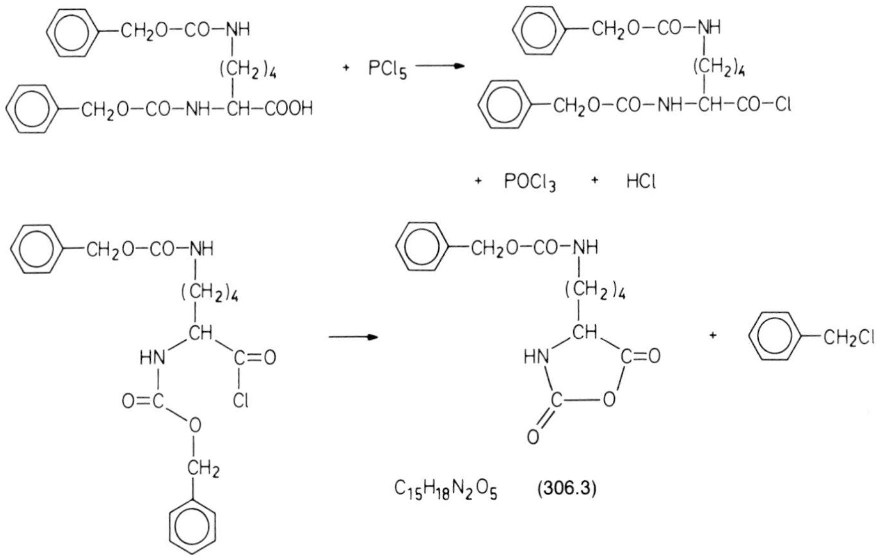

$C_{15}H_{18}N_2O_5$ (306.3)

A solution of L-lysine dihydrochloride (22 g, 100 mmol) in 2 N NaOH (150 ml) is cooled in an ice-water bath and vigorously stirred. Benzyl chlorocarbonate [3] (53 g = 44 ml, ca. 300 mmol) and 4 N NaOH (125 ml) are added, each in four portions. These additions require about 30 min. Stirring is continued for an additional 30 min then the mixture [4] is acidified to congo with 6 N hydrochloric acid (about 55–60 ml). Di-benzyloxycarbonyl-L-lysine separates as an oil which is extracted into ether (twice, 200 ml each time). The organic layers are combined and extracted with a 5% solution of $KHCO_3$ in water (three times, 200 ml each time) [5]. Acidification of the combined bicarbonate solutions with 6 N HCl (60 ml) and reextraction with ether (twice, 150 ml each time), drying the ether extracts over anhydrous Na_2SO_4 and evaporation of the solvent under reduced pressure leaves di-benzyloxycarbonyl-L-lysine (40 g) as a syrup, which is used without further purification [6].

The syrup is dissolved in dry ether (150 ml) and the solution cooled in an ice-water bath. Powdered phosphorus pentachloride (23 g, 110 mmol) is added and the mixture shaken at about 10 °C until most of the PCl_5 disappears. This requires about 30 min. The solution is decanted from the small amount of insoluble material and rapidly concentrated in vacuo. An oil bath of about 50 °C is used and moisture is carefully excluded during these operations. Dry ethyl acetate (100 ml) is added to the residue and removed in vacuo. The addition of ethyl acetate (100 ml) and its removal are repeated and the residue crystallized from a small volume of ethyl acetate by dilution with hexane. The *N*-carboxyanhydride (24.5 g, 80%) melts at 100 °C with decomposition [7].

1. *N*-carboxyanhydrides (4-substituted oxazolidine-2,4-diones) are also called Leuchs anhydrides; cf. Leuchs H, Geiger W (1908) Ber dtsch Chem Ges 41: 1721; Wessely F (1925) Hoppe Seyler's Z Physiol Chem 146: 72. While they are generally prepared through the reaction of amino acids with phosgene, the method described in this section is quite convenient.
2. Bergmann M, Zervas L, Ross WF (1935) J Biol Chem 111: 245 (1935)
3. Benzyl chlorocarbonate is commercially available but mostly under the name of benzyl chloroformate or carbobenzoxy chloride. It is an irritant and, therefore, all the operations described here should be carried out under a well-ventilated hood.
4. The sodium salt of di-benzyloxycarbonyl-L-lysine might form an oil.
5. During this extraction CO_2 evolves.
6. Pure di-benzyloxycarbonyl-L-lysine is crystalline and melts at 150 °C. For the preparation of the *N*-carboxyanhydride the syrup is preferable because it is soluble in ether.
7. On storage for longer periods of time the m.p. can reach 250 °C or higher, but this is probably due to polymerization. *N*-Carboxyanhydrides can be stored but only with the rigorous exclusion of moisture.

4.2 Poly-N^{ε}-benzyloxycarbonyl-L-lysine [1, 2]

C_6H_5–CH_2O–CO–NH–$(CH_2)_4$–CH(HN–C(=O)–O–C=O) $\xrightarrow[\text{HOH (trace)}]{\Delta}$ H(–NH–CH[$(CH_2)_4$–NH–CO–OCH_2–C_6H_5]–CO–$)_n$OH

N^{ε}-benzyloxycarbonyl-N^{α}-carboxy-L-lysine anhydride is recrystallized 6 times from dry ethyl acetate-petroleum ether and immediately introduced into a dry glass apparatus connected, through a trap containing P_2O_5 and a second trap cooled with liquid air, to a high vacuum source. The material is dried at 60 °C and 10^{-4} mm for several hours. The temperature is raised to 105 °C: the anhydride melts and evolution of CO_2 can be observed. After about an hour gas evolution ceases and the reaction vessel contains the polymer, a transparent, glassy material. It is insoluble in water and in ether but soluble in hot acetic acid and to some extent also in hot ethanol. From a solution in hot acetic acid a part of the polymer separates on cooling and more is precipitated on dilution with water. The data of elemental analysis correspond to a polymer with an average polymerization number of 32.

1. Katchalski E, Grossfeld I, Frankel M (1948) J Amer Chem Soc 70: 2094; Cf. also the preceding Sect. 4.1
2. For the preparation of poly-L-aspartic acid cf. Berger A, Katchalski E (1951) J Amer Chem Soc 73: 4084

4.3 Synthesis of Sequential Peptides by Polycondensation

Poly-(L-lysyl-L-alanyl-L-alanine) Trifluoroacetate [1]

To a stirred solution of L-alanine 2-benzyloxyphenyl ester hydrochloride [1, 2] (3.1 g, 10 mmol) in dimethylformamide (20 ml) triethylamine (1.01 g = 1.4 ml, 10 mmol) and *tert*-butyloxycarbonyl-L-alanine *N*-hydroxysuccinimide ester [3] (2.86 g, 10 mmol) are added. The mixture is stirred at room temperature for 5 hours, then diluted with ethyl acetate (250 ml) and water (80 ml). The two layers are separated, the ethyl acetate solution washed with a 10% solution of citric acid in water (100 ml), with a saturated aqueous solution of $NaHCO_3$ (100 ml) and with water (100 ml), dried over anhydrous $MgSO_4$ and evaporated in vacuo. The residue, an oil, is triturated with petroleum ether (b.p. 40–60 °C), the solid collected on a filter, washed with petroleum ether and dried in air. Recrystallization from ether-petroleum ether affords *tert*-butyloxycarbonyl-L-alanyl-L-alanine 2-benzyloxyphenyl ester in the form of white needles (3.8 g, 86%) melting at 100–120 °C; $[\alpha]_D^{20}$ −49.6 (*c* 1, $CHCl_3$). The protected dipeptide ester appears as a single spot on thin layer chromatograms.

The protected dipeptide ester (2.21 g, 5 mol) is dissolved in 90% aqueous trifluoroacetic acid (7 ml) and the solution kept at room temperature for one hour. The volatile materials are removed in vacuo and the residue dried first by repeated evaporations with benzene then in vacuo (ca 10 Pa) at room temperature for several hours. The dry trifluoroacetate salt, a glass, is dissolved in dimethylformamide (7 ml) and treated with triethylamine (0.51 g = 0.70 ml, 5 mmol) and with N^α-benzyloxycarbonyl-N^ε-*tert*-butyloxycarbonyl-L-lysine *N*-hydroxysuccinimide ester [4] (2.4 g, 5 mmol). The reaction is allowed to proceed at room temperature overnight. Ethyl acetate (250 ml) and water (80 ml) are added, the phases separated and the organic layer washed with a 10% aqueous citric acid solution (80 ml), saturated $NaHCO_3$ solution (80 ml), and with water (80 ml), dried over anhydrous $MgSO_4$ and evaporated in vacuo. The solid residue, the protected tripeptide ester, is recrystallized from chloroform-petroleum ether. The purified preparation (needles, 2.75 g, 78%) melts at 164–166 °C; $[\alpha]_D^{20}$ −55.7 ° (*c* 1, $CHCl_3$), is homogeneous on thin layer chromatograms and gives satisfactory C, H and N values on elemental analysis.

$$(CH_3)_3C{-}O{-}CO{-}NH{-}CH(CH_3){-}CO{-}O{-}N(CO)_2C_2H_4 \;+\; H_2N{-}CH(CH_3){-}CO{-}O{-}C_6H_4{-}O{-}CH_2{-}C_6H_5 \longrightarrow$$

$$(CH_3)_3C{-}O{-}CO{-}NH{-}CH(CH_3){-}CO{-}NH{-}CH(CH_3){-}CO{-}O{-}C_6H_4{-}O{-}CH_2{-}C_6H_5 \xrightarrow{CF_3COOH}$$

$$CF_3COOH\cdot H_2N{-}CH(CH_3){-}CO{-}NH{-}CH(CH_3){-}CO{-}O{-}C_6H_4{-}O{-}CH_2{-}C_6H_5 \xrightarrow{C_6H_5{-}CH_2{-}O{-}CO{-}NH{-}CH[(CH_2)_4{-}NH{-}CO{-}O{-}C(CH_3)_3]{-}CO{-}O{-}N(CO)_2C_2H_4}$$

$$C_6H_5{-}CH_2O{-}CO{-}NH{-}CH[(CH_2)_4{-}NH{-}CO{-}O{-}C(CH_3)_3]{-}CO{-}NH{-}CH(CH_3){-}CO{-}NH{-}CH(CH_3){-}CO{-}O{-}C_6H_4{-}O{-}CH_2{-}C_6H_5 \xrightarrow[(AcOH)]{H_2/Pd}$$

$$\left[H_2N{-}CH[(CH_2)_4{-}NH{-}CO{-}O{-}C(CH_3)_3]{-}CO{-}NH{-}CH(CH_3){-}CO{-}NH{-}CH(CH_3){-}CO{-}O{-}C_6H_4{-}OH\right] \longrightarrow$$

$$H_2NCH[(CH_2)_4NH_2{-}Boc]CO{-}NHCH(CH_3)CO{-}NHCH(CH_3)CO{-}(NHCH[(CH_2)_4NH_2{-}Boc]CO{-}NHCH(CH_3)CO{-}NHCH(CH_3)CO{-})_n\,NHCH[(CH_2)_4NH_2{-}Boc]CO{-}NHCH(CH_3)CO{-}NHCH(CH_3)COOH$$

$$\xrightarrow{CF_3COOH} H_2NCH[(CH_2)_4NH_2]CO{-}NHCH(CH_3)CO{-}NHCH(CH_3)CO{-}(NHCH[(CH_2)_4NH_2]CO{-}NHCH(CH_3)CO{-}NHCH(CH_3)CO{-})_n\,NHCH[(CH_2)_4NH_2]CO{-}NHCH(CH_3)CO{-}NHCH(CH_3)COOH\cdot(n{+}1)\,CF_3COOH$$

The fully protected tripeptide derivative, N^{α}-benzyloxycarboxyl-N^{ε}-*tert*-butyloxycarbonyl-L-lysyl-L-alanyl-L-alanine 2-benzyloxyphenyl ester, (1.41 g, 2 mmol) is dissolved in acetic acid (30 ml) and after the addition of a 10% Pd on charcoal catalyst (1.4 g) hydrogenated at room temperature and atmospheric pressure for about 3 hours. The catalyst is removed by filtration through a layer of Celite and the solvent evaporated in vacuo. The residue is dried in a good vacuum (about 10 Pa) at room temperature. The dry material, a foam, is dissolved in dimethylsulfoxide (2 ml) and the stirred solution is treated with triethylamine (0.41 g = 0.56 ml, 4 mmol). The mixture is allowed to stand at room temperature. Four days later the reaction mixture, a semisolid mass, is triturated with 95% ethanol (20 ml, in several portions), the precipitate collected by centrifugation and washed with 95% ethanol (40 ml, in several portions) and similarly with ether (100 ml). For final deprotection the air-dry material is dissolved in 90% aqueous trifluoroacetic acid and kept at room temperature for one hour. Ether (30 ml) is added, the precipitate collected by centrifugation, washed with ether (120 ml, in several portions) and dried. The crude polymer (trifluoroacetate salt, about 0.28 g) is dissolved in water (25 ml) and dialyzed against distilled water (5 liter) for about 24 h, with the water changed four times. The purified polymer is recovered by removal of the water from the frozen solution (lyophilization). The polymer is dried in a good vacuum, (about 10 Pa) at room temperature. It weighs about 116 mg (15%) and melts, with decomposition at 175–190 °C; $[\alpha]_D^{20}$ $-89°$ (*c* 1, buffer of pH 7.0). A 0.63% solution of the polymer in dichloroacetic acid has reduced viscosity (ηsp/c) of 0.265 dl g^{-1}. The values of elemental analysis agree with those calculated for $C_{14}H_{23}F_3N_4O_5 \cdot 1H_2O$.

1. Cowell RD, Jones JH, J Chem Soc Perkin I: 1972, 2236
2. Described in this volume on page 44
3. Anderson GW, Zimmermann JE, Callahan FM (1964) J Amer Chem Soc 86: 1839
4. Prepared according to a published procedure (Otsuka H, Inouye K, Kanayama M, Shinozaki F (1966) Bull Chem Soc Japan 39: 882 in a 1:1 mixture of ethyl acetate and dioxane at 0 °C. Recrystallization from 2-propanol–light petroleum ether gave the active ester in 72% yield with m.p. 98–99 °C; $[\alpha]_D^{20}$ $-16.8°$ (*c* 1.7, dioxane).

5 Partial Synthesis

Human Insulin [1, 2]

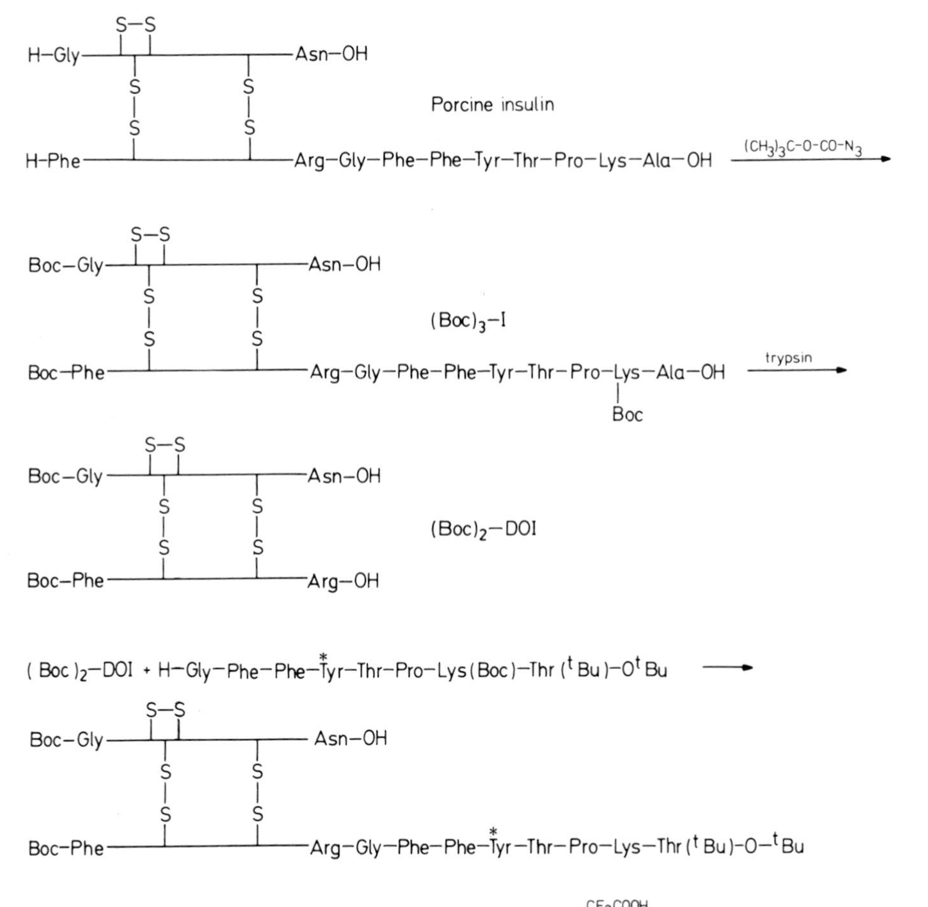

Zinc Free Insulin [3]

Zinc insulin [4] (1.2 g, 0.2 mmol) is dissolved in 0.025 N HCl and dialyzed at 0 °C for 5 days [5], with frequent changes of the 0.025 N HCl. Lyophilization of the content of the dialysis bag leaves insulin hydrochloride (1.2 g) which has a zinc content of about 0.04% [6].

---

* In the procedure described in ref. 1 the phenolic hydroxyl group of the tyrosine residue is blocked in the form of tert. butyl ether, in ref. 2 no protection of this function is mentioned.

N^{A1},-N^{B1}-Ditert-butyloxy-carbonyl-B(23–30)-desoctapeptide Insulin, $(Boc)_2$-DOI

The hydrochloride of pork insulin (0.60 g, 0.1 mmol) is dissolved in purified dimethylformamide [7] (80 ml). Triethylamine (0.13 g = 0.18 ml, 1.3 mmol) and *tert*-butyl azidoformate [8] (5.2 g, 36 mmol) are added and the mixture stirred at 40 °C for about 5 hours. It is cooled to room temperature and diluted with ether until no more precipitate forms. The solid material is collected by centrifugation, washed with ether and dried. The crude N^{A1},-N^{B1}, N^{B30}, -tri-*tert*-butyloxycarbonyl porcine insulin [9], $(Boc)_3$-I, (about 0.60 g) thus obtained is used directly for the preparation of the desoctapeptide derivative.

The tri-Boc derivative (0.63 g, 0.1 mmol) is dissolved in dimethylformamide [7] (15 ml) and treated with a 0.05 molar tris buffer [10] of pH 7.5 (9 ml), 10^{-3} molar with respect to $CaCl_2$. A solution of trypsin (32 mg) in the same buffer (1.0 ml) is added at hourly intervals in ten portions while the temperature of the solution is maintained at 35 °C. The reaction mixture is stored overnight at room temperature then its pH adjusted [11] to 4.5 with 0.01 N HCl. The solvents are removed in vacuo, the residue triturated with acetone, washed with ether and dried in air. The peptides are separated by partition chromatography on a Sephadex-LH20 column in the system *n*-butanol-acetic acid-water (4:1:10). About 0.3 g of $(Boc)_2$-DOI is obtained.

Human Insulin Through Condensation with Dicyclohexyl-carbodiimide [1]

$(Boc)_2$-DOI (506 mg, 0.1 mmol) together with *N*-methylmorpholine (61 mg = 0.067 ml, 0.60 mmol), 1-hydroxybenzotriazole (15 mg, 0.1 mmol) and the partially protected octapeptide *H*-Gly-Phe-Phe-Tyr(^{t}Bu)-Thr-Pro-Lys(Boc)-Thr(^{t}Bu)-O^{t}Bu [12] (130 mg, 0.1 mmol) are added to dimethylformamide [7] (6 ml) and the solution is treated with dicyclohexylcarbodiimide (21 mg, 0.1 mmol). The reaction mixture is stored at room temperature overnight. The pH of the solution is adjusted to 4.5 and the peptidic material precipitated by dilution with ether. The product is triturated with acetone, washed with ether and dried in air (about 530 mg). The crude mixture is dissolved in trifluoroacetic acid (4 ml) containing anisole (0.20 ml). After one hour at room temperature the solution is cooled in an ice-water bath and diluted with ether to precipitate the crude product. The latter is collected on a filter, thoroughly washed with ether and dried in vacuo over KOH pellets. The preparation [13] thus obtained shows by bioassay an insulin content of 25 to 40%; pure, crystalline human insulin can be obtained from it by chromatography.

Human Insulin via Enzyme-Catalyzed Coupling [2]

The partially blocked desoctapeptide insulin $(Boc)_2$-DOI (100 mg) and the octapeptide derivative *H*-Gly-Phe-Phe-Tyr-Thr-Pro-Lys-Thr(^{t}Bu)-*O*^{t}Bu [14] (200 mg, ten-fold excess over the desoctapeptide-insulin derivative) are dissolved in a mixture of dimethylformamide [7] (0.60 ml) and 0.25 molar tris-buffer (0.60 ml). The solution is kept at 37 °C and treated with trypsin [15] (3.3 mg). The same amount of the enzyme is added after 2 h and again after 6 h. Formation of the peptide bond can be monitored by liquid chromatography [2] which indicates 58% conversion after 20 h. The entire mixture is applied to a column of Sephadex LH-20 and the peptides eluted with a 1:1 mixture of

DMF [7] and 0.5 molar acetic acid. Unreacted $(Boc)_2$-DOI and the newly formed insulin are eluted in a single peak which is followed by a second peak that corresponds to the octapeptide derivative applied in excess. Most of the latter can be recovered by rechromatography.

The material in the first peak is precipitated with ether, washed with ether, dried and dissolved in ice-cold trifluoroacetic acid (2 ml) containing anisole (0.10 ml). After one hour at 0 °C the peptides are precipitated with ether and chromatographed on a Sephadex 50-superfine column with 0.5 molar acetic acid as eluent. Human insulin (51 mg) is obtained from one peak and DOI (37 mg) from a second [16]. Rechromatography of the overlapping portion of the two peaks affords more human insulin (5 mg) and raises the total yield to 56 mg (49%). The semisynthetic material gives satisfactory amino acid analysis and according to electrophoresis on polyacrylamide gel it is as pure as the crystalline pork insulin used as starting material.

1. Obermeier R, Geiger R (1976) Hoppe Seyler's Z Physiol Chem 357: 759
2. Inouye K, Watanabe K, Morihara K, Tocino Y, Kanaya T, Emura J, Sakakibara S (1979) J Amer Chem Soc 101: 751
3. Laskowski M, Jr, Leach SJ, Sheraga H (1960) J Amer Chem Soc 82: 571
4. In porcine insulin, and also in the insulin of whales, the C-terminal residue of the B-chain is alanine while in human insulin this position is occupied by threonine. The insulin molecules of many other species are different at several positions. Thus, pork or whale insulin must be used for the preparation of human insulin by semisynthesis.
5. A rocking dialyzer is recommended in ref. 3.
6. In zinc insulin one zinc atom is coordinated with two molecules of insulin. The zinc content of the complex is 0.57%.
7. Dimethylformamide is purified by filtration through a column of neutral alumina.
8. Freshly prepared from *tert*-butyl carbazate (cf. p. 195) with nitrous acid.
9. Levy D, Carpenter FH (1967) Biochemistry 6: 3559
10. Tris(hydroxymethyl)aminomethane.
11. The pH values in this section are meant as readings on a glass electrode.
12. The octapeptide derivative was synthesized by conventional means.
13. A characteristic feature of the partial synthesis of human insulin from $(BOc)_2$-DOI is that six free carboxyl groups compete in the reaction with the activating reagent (dicyclohexylcarbodiimide). While the desired reaction, activation of the C-terminal residue of the B-chain, an arginine moiety, occurs preferentially, a considerable number of by-products are also present in the crude synthetic preparation.
14. No details on the preparation of this partially protected octapeptide are described in ref. 2.
15. A trypsin preparation treated with the chymotrypsin inhibitor L-(1-tosylamido-2-phenyl)ethyl chloromethyl ketone (TPCK) was used; cf. Kostka V, Carpenter FH (1964) J Biol Chem 239: 1799.
16. Instead of $(Boc)_3$-I this material (DOI) can also be used for the preparation of $(Boc)_2$-DOI. Tryptic digestion of insulin yields DOI; cf. Bromer WW, Chance EC (1967) Biochem Biophys Acta 133: 219 (1967). Treatment of DOI with *tert*-butyl 1-succinimidyl carbonate converts DOI into $(Boc)_2$-DOI (cf. ref. 1).

VI Models for the Study of Racemization

1 Benzoyl-leucyl-glycine Ethyl Ester [1, 2]

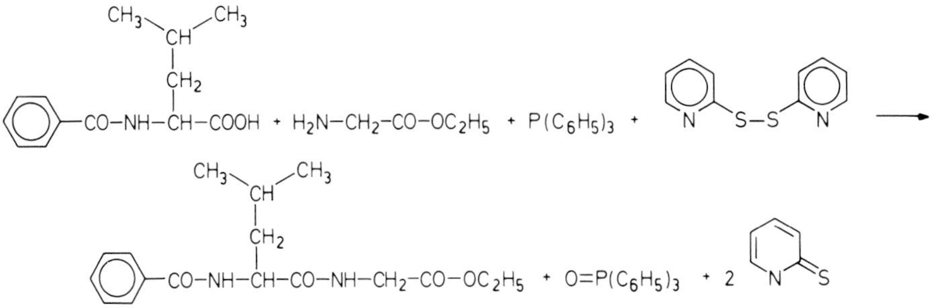

C_6H_5–CO–NH–CH(CH_2–CH(CH_3)$_2$)–COOH + H_2N–CH_2–CO–OC_2H_5 + $P(C_6H_5)_3$ + 2,2′-dipyridyldisulfide (N ... S–S ... N) →

C_6H_5–CO–NH–CH(CH_2–CH(CH_3)$_2$)–CO–NH–CH_2–CO–OC_2H_5 + $O{=}P(C_6H_5)_3$ + 2 pyridine-2-thione (N ... S)

$C_{17}H_{24}N_2O_4$ (320.4)

A solution of 2,2′-dipyridyldisulfide [3] (2.20 g, 10 mmol) and benzoyl-L-leucine [1] (2.35 g, 10 mmol) in dichloromethane (60 ml) is stirred at room temperature while a solution of triphenylphosphine (2.62 g, 10 mmol) and glycine ethyl ester (1.04 g, 10 mmol) is added dropwise. After the addition is complete stirring is continued for 30 min. The solution is washed with N HCl (50 ml), water (50 ml), a 5% solution of $NaHCO_3$ in water (50 ml) and water (50 ml). It is dried over anhydrous Na_2SO_4 and evaporated to dryness in vacuo. The residue is triturated with light petroleum ether (b.p. 30–50 °C), the solid product filtered, washed with petroleum ether and dried. The benzoyl-dipeptide ester (2.92 g, 91%) melts at 148–152 °C; $[\alpha]_D^{20}$ −32.6° (*c* 3.1, ethanol). This value of specific rotation corresponds [1] to a mixture of 96% L and 4% D isomer.

1. Synthesis of benzoyl-L-leucyl-glycine ethyl ester is a test of various coupling methods and conditions of coupling with respect to their influence on racemization (Williams MW, Young GT, J Chem Soc 1963: 881).
2. Application of the Young test is shown here on the example of the oxidation-reduction method of coupling (Matsueda R, Maruyama H, Ueki M, Mukayama T (1971) Bull Chem Soc Jpn 44: 1373). The favorable outcome of the cited experiment should be regarded with certain caution. For instance, the extent of racemization could be quite different if instead of dichloromethane dimethyl-formamide had been used. Also, the application of an amine salt together with a tertiary amine as amino component usually gives less good results than the use of a free amine such as glycine ethyl ester in the above experiment.
3. Available commercially as 2,2′-dithiodipyridine.

2 The Anderson-Callahan Test [1]

Benzyloxy-carbonyl-glycyl-L-phenylalanyl-glycine Ethyl Ester [2]

$$C_6H_5-CH_2O-CO-NH-CH_2-CO-NH-CH(CH_2C_6H_5)-COOH + H_2N-CH_2-CO-OC_2H_5 + (C_3H_3N_2)-CO-(C_3H_3N_2)$$

$$\longrightarrow C_6H_5-CH_2O-CO-NH-CH_2-CO-NH-CH(CH_2C_6H_5)-CO-NH-CH_2-CO-OC_2H_5 + 2\,HN(C_3H_3N) + CO_2$$

$C_{23}H_{27}N_3O_6$ (441.5)

Benzyloxycarbonyl-glycyl-L-phenylalanine [3] (3.56 g, 10 mmol) is dissolved in dry [4] dimethylformamide (10 ml). The solution is cooled to −10 °C and carbonyldiimidazole [5] (1.65 g, 10 mmol of a material of 98% purity) is added; a slow evolution of CO_2 can be observed. When the effervescence stops, glycine ethyl ester [6] (1.03 g, 10 mmol) is added and the reaction mixture is allowed to warm up to room temperature. After about 30 min at room temperature, N HCl (50 ml) is added. An oil separates and gradually solidifies. It is disintegrated, collected on a filter, washed with a 5% solution of $NaHCO_3$ in water (20 ml), then with water (30 ml in three portions) and dried over P_2O_5 in vacuo. The protected tripeptide derivative (4.22 g, 96%) [7] melts at 115.5–117 °C. It is dissolved in absolute ethanol (210 ml) to yield a 2% solution. The latter is cooled in an ice-water bath and seeded with the racemic form of the protected tripeptide. Crystallization is interrupted from time to time, the crystals are collected and examined by weight and melting point. In the example cited here the following fractions were obtained: 9 mg (m.p. 120–133.5 °C); 10 mg (m.p. 119–128.5 °C; 21 mg (m.p. 118.5 to 119.5 °C); 2.32 g (120–120.5 °C). After concentration to a small volume a further fraction weighing 1.48 g (m.p. 120–120.5 °C) was secured. Complete removal of the solvent from the filtrate left a residue (0.33 g). Since the pure racemate melts at 132–133 °C, the melting points reveal a small amount of racemate in the first two fractions. The total weight of these two fractions (19 mg) and the total amount recovered in crystalline form (3.8 g) indicate less than 0.5% racemic material [8].

1. Anderson GW, Callahan FM (1960) J Amer Chem Soc 82: 3359
2. Paul R, Anderson GW (1960) J Amer Chem Soc 82: 4596. The racemization test is demonstrated here on the example of coupling with the help of carbonyldiimidazole. Of course, the synthesis of this model compound is suitable for the study of other methods of coupling and conditions of coupling as well.
3. An enantiomerically homogeneous preparation with m.p. 127.5–128 °C and $[\alpha]_D^{25}$ +38.2 (*c* 5, abs. ethanol) should be used. Anderson GW, Callahan FM (1958) J Amer Chem Soc 80: 2902
4. The solvent is dried over calcium hydride. The operations should be carried out under the exclusion of moisture: carbonyldiimidazole readily reacts with water.
5. Commercially available.
6. Freshly distilled.
7. Since the test is based on negative evidence (the lack of separation of the fairly insoluble racemate), it is valid only if no significant amounts of by-products are present in the crude material.
8. Unless the residue (0.33 g) contains an appreciable amount of benzyloxycarbonylglycyl-DL-phenylalanyl-glycine ethyl ester.

3 Acylation with Acetyl-L-isoleucine [1]

Acetyl-L-isoleucine [2] An enantiomerically pure sample of L-isoleucine [3] (13.1 g, 100 mmol) is added to ice-cold N NaOH (100 ml) and the stirred solution is cooled in an ice-water bath. Chilled N NaOH (20 ml) is added followed by acetic anhydride (2.04 g = 0.94 ml, 20 mmol). The anhydride soon disappears and a homogeneous solution forms. The reaction of the solution should be distinctly alkaline; if necessary a small volume of N NaOH is added to assure alkalinity. The addition of N NaOH and acetic anhydride (the same amounts) is repeated four more times with testing for alkalinity and adjustment, if necessary, after each addition. Stirring is then continued for 30 min. The solution is acidified to Congo with concentrated hydrochloric acid (about 8.5 ml) and crystallization is allowed to proceed at 0 °C for one hour. The crystals are collected on a filter, washed with ice-cold water (25 ml) and dried in air. The crude material (about 8 g) melts at 148–150 °C. Recrystallization from boiling water (80 ml) raises the m.p. to 149–151 °C; concentration of the mother liquor provides a second crop with similar melting point [4].

Acetyl-L-isoleucyl-glycine Ethyl Ester and Acetyl-D-alloisoleucyl-glycine Ethyl Ester

$CH_3CO{-}NH{-}CH(CH(CH_3)CH_2CH_3){-}COOH + H_2N{-}CH_2{-}CO{-}OC_2H_5 +$ C₆H₁₁–N=C=N–C₆H₁₁ $\longrightarrow$

$CH_3CO{-}NH{-}CH(CH(CH_3)C_2H_5){-}CO{-}NH{-}CH_2{-}CO{-}OC_2H_5 +$ C₆H₁₁–NH–C(=O)–NH–C₆H₁₁

$C_{12}H_{22}N_2O_4$ (258.3)

$\xrightarrow{HCl/HOH} CH_3COOH + HCl\cdot H_2N{-}CH(CH(CH_3)C_2H_5){-}COOH + HCl\cdot H_2N{-}CH_2COOH \xrightarrow{\text{ion-exchange chromatography}}$

$H_2N{-}CH_2{-}COOH$ — glycine

L-isoleucine (Fischer projection: COOH; H_2N–C–H; CH_3–C–H; C_2H_5)

D-alloisoleucine (Fischer projection: COOH; H–C–NH_2; CH_3–C–H; C_2H_5)

A sample of acetyl-L-isoleucine (173 mg, 1 mmol) is added to a solution of glycine ethyl ester hydrochloride (173 mg, 1.25 mmol) and triethylamine [5] (126 mg = 0.175 ml, 1.25 mmol) in chloroform [5] (5 ml). The mixture is cooled in an ice-water bath, treated with dicyclohexylcarbodiimide [5] (206 mg, 1 mmol) and kept in the ice-water bath for one hour and at room temperature for an additional hour. One drop of acetic acid is added and the *N,N'*-dicyclohexylurea which separated during the reaction is removed by filtration [6] and washed with chloroform (5 ml). The combined filtrate and washings are diluted with chloroform (20 ml) and extracted with 0.5 N $KHCO_3$ (20 ml), water (20 ml). 0.5 N HCl (20 ml) and water (20 ml), dried over $MgSO_4$, filtered into a weighed round bottom flask and evaporated to dryness in vacuo. The residue is dried to constant weight in the vacuum of an oil pump at room temperature [7].

A small aliquot (about 2.6 mg, accurately weighed) of the crude coupling product is used for the determination of racemization. The sample is weighed into an ampoule and dissolved in distilled constant boiling hydrochloric acid (0.5 to 1.0 ml). The solution is cooled in an ice-water bath, evacuated with the help of a water aspirator and sealed. Hydrolysis is carried out at 110 °C for 16 hours [8]. The sample is then cooled to room temperature, the ampoule opened and its content quantitatively transferred into a 50 ml beaker and evaporated to dryness on the steam bath. Distilled water (about 2 ml) is added and the evaporation repeated. The residue is dissolved in a 0.2 M citrate buffer of pH 2.2 (10.0 ml) and a suitable aliquot [9] is applied to the column of an amino acid analyzer. The amount of glycine found on analysis should be equal with the sum of the amounts of isoleucine and alloisoleucine present in the mixture [10]. The amount of alloisoleucine divided by the sum of the amounts is isoleucine and alloisoleucine and multiplied by 100 gives the extent of racemization in percents [11].

1. Bodanszky M, Conklin LE, Chem Communications 1967: 773
2. Greenstein JP, Winitz M (1961) The Chemistry of Amino Acids, John Wiley, New York, p. 2065
3. The purity of the sample is best checked with the help of an amino acid analyzer. With a buffer of pH 4.25 alloisoleucine is eluted after methionine and before isoleucine in a peak well separated from the peaks of both these amino acids. The sample should contain less than 0.5% alloisoleucine. The specific rotation of pure L-isoleucine is $[\alpha]_D^{25} + 39.6°$ (*c* 1, 5 N HCl) and +48.8° (*c* 1, acetic acid).
4. Drying at elevated temperatures might result in losses. Acetyl-L-isoleucine sublimes unchanged at 90 °C and about 10 Pa.
5. The test is suitable for the determination of the effect of the solvent, tertiary amine, temperature, additives, etc. on the extent of racemization during coupling and of the influence of the method of coupling itself. Thus coupling methods and conditions other than here described can be applied. An important feature of the test is that it does not involve purification and hence the ratio of the L-isoleucine derivative to the alloisoleucine containing dipeptide is not modified in the process.
6. The amount of the precipitate (calculated 224 mg) and its melting point provide important information on the outcome of the coupling reaction. *N,N'*-dicyclohexylurea changes

crystal form at 218 °C and melts between 228 and 232 °C. Of course, the weight of the crude product is equally important in this respect.

7. At higher temperature the acetyldipeptide ester sublimes in vacuo.
8. An electrically heated block is best suited for this purpose.
9. The size of the sample depends on the sensitivity of the analyzer.
10. A hydrolysate of the starting material acetyl-L-isoleucine also contains some alloisoleucine, formed during hydrolysis. This value should be applied as a correction in the determination of the amount of alloisoleucine produced in the coupling reaction.
11. While it is customary to express racemization as the fraction of alloisoleucine formed, in reality twice this figure corresponds to the part of the material participating in the process: half of the molecules which loose chirality are reconverted to L-isoleucine residues. Thus, the true extent of racemization is twice the value calculated in the above indicated manner.

4 The Izumiya-Muraoka Model

Glycyl-alanyl-leucine [1]

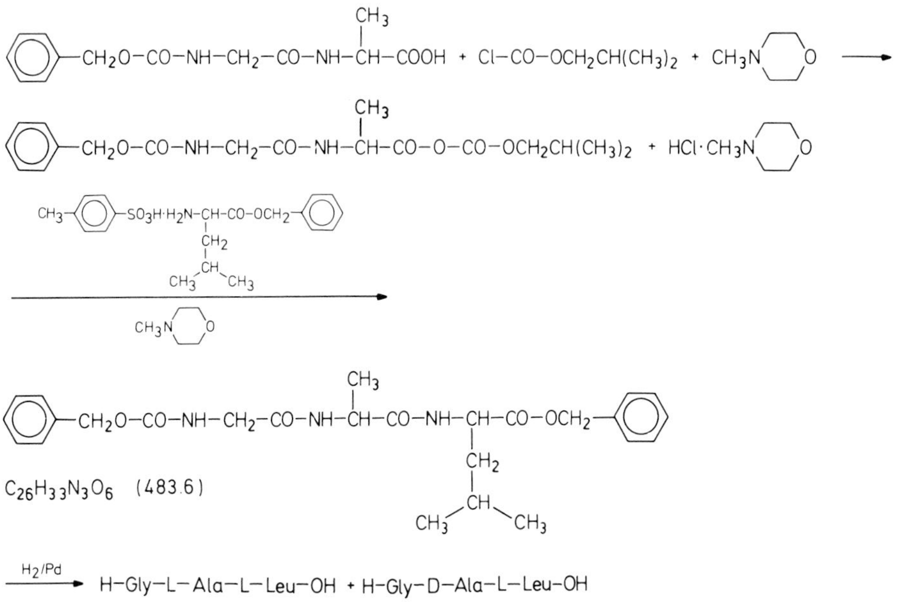

$C_6H_5-CH_2O-CO-NH-CH_2-CO-NH-CH(CH_3)-COOH + Cl-CO-OCH_2CH(CH_3)_2 + CH_3N(C_4H_8O) \longrightarrow$

$C_6H_5-CH_2O-CO-NH-CH_2-CO-NH-CH(CH_3)-CO-O-CO-OCH_2CH(CH_3)_2 + HCl \cdot CH_3N(C_4H_8O)$

$\xrightarrow[CH_3N(C_4H_8O)]{CH_3-C_6H_4-SO_3H \cdot H_2N-CH(CH_2-CH(CH_3)_2)-CO-OCH_2-C_6H_5}$

$C_6H_5-CH_2O-CO-NH-CH_2-CO-NH-CH(CH_3)-CO-NH-CH(CH_2-CH(CH_3)_2)-CO-OCH_2-C_6H_5$

$C_{26}H_{33}N_3O_6$ (483.6)

$\xrightarrow{H_2/Pd}$ H–Gly–L–Ala–L–Leu–OH + H–Gly–D–Ala–L–Leu–OH

A solution of benzyloxycarbonyl-glycyl-L-alanine [2] (2.80 g, 10 mmol) in dry tetrahydrofurane (50 ml) is cooled to $-15\,^{\circ}C$ and *N*-methylmorpholine (1.01 g $=$ 1.10 ml, 10 mmol) is added, followed by the addition of isobutyl chlorocarbonate (1.37 g $=$ 1.32 ml, 10 mmol). The mixture is stirred at $-15\,^{\circ}C$ for 12 minutes then L-leucine benzyl ester *p*-toluenesulfonate [3] (3.94 g, 10 mmol) and *N*-methylmorpholine (1.01 g $=$ 1.10 ml, 10 mmol) are added and the mixture allowed to warm up to room temperature. Fifteen hours later the solvent is removed under reduced pressure, the residue dissolved in a mixture of ethyl acetate (150 ml) and a 5% solution of $NaHCO_3$ in water (50 ml). The organic layer is extracted with water (50 ml), N HCl (50 ml) and water (50 ml), dried over anhydrous Na_2SO_4 and evaporated to dryness in vacuo. The crude tripeptide derivative weighs about 5 g; an aliquot of this material is used without purification [4] for the determination of racemization.

A 10% aliquot (about 0.5 g) of the crude material is dissolved in a mixture of acetic acid (18 ml) and water (2 ml), a 10% palladium on charcoal catalyst (0.20 g) is added and after the displacement of air with nitrogen the mixture is stirred in a hydrogen atmosphere for 6 hours. The hydrogen is displaced with nitrogen, the catalyst is removed by filtration and the filtrate evaporated to dryness in vacuo. The residue is dissolved in a 0.2 molar citrate buffer of pH 4.25 (100 ml) and an aliquot [5] of this solution is applied on the long column of an amino acid analyzer [6]. The peak of *H*-Gly-L-Ala-L-Leu-OH is followed by that of *H*-Gly-D-Ala-L-Leu-OH and a determination of the areas under these peaks allows the calculation of racemization that took place in the coupling reaction. In the above example a 91% yield was found [7] and 2.4% of the diastereomeric mixture consisted of the tripeptide which contained the D-alanine residue [8].

1. Izumiya N, Muraoka M (1969) J Amer Chem Soc 91: 2391
2. Erlanger BF, Brand E (1951) J Amer Chem Soc 73: 3508
3. The preparation of amino acid benzyl ester *p*-toluenesulfonates is described on page 30.
4. Purification might distort the picture since the ratio of the two diastereoisomers in the purified material is probably different from their ratio in the crude product.
5. The amount used for amino acid analysis depends on the type of the available analyzer.
6. In the here cited experiment a Hitachi Model KLA-3B instrument was used with spherical Dowex 50 resin in a 0.9 × 50 cm column. The L-L-tripeptide appeared at an elution volume of 129 ml while the D-L-tripeptide emerged at 159 ml. For other analyzer types the elution volume of the two isomers has to be established with the aid of authentic samples.
7. The overall yield is calculated from the recovery of the amino acid analysis.
8. Under the condition described above but with triethylamine instead of *N*-methylmorpholine the overall yield was 87% and the mixture contained 9.5% of the undesired isomer, *H*-Gly-D-Ala-L-Leu-OH. Synthesis of these model peptides allows the study of the influence on racemization of solvents, tertiary amines, temperature and other conditions of the coupling and last but not least the method of coupling itself.

5 Synthesis and Enantioselective Enzymic Hydrolysis of Tetraalanine [1]

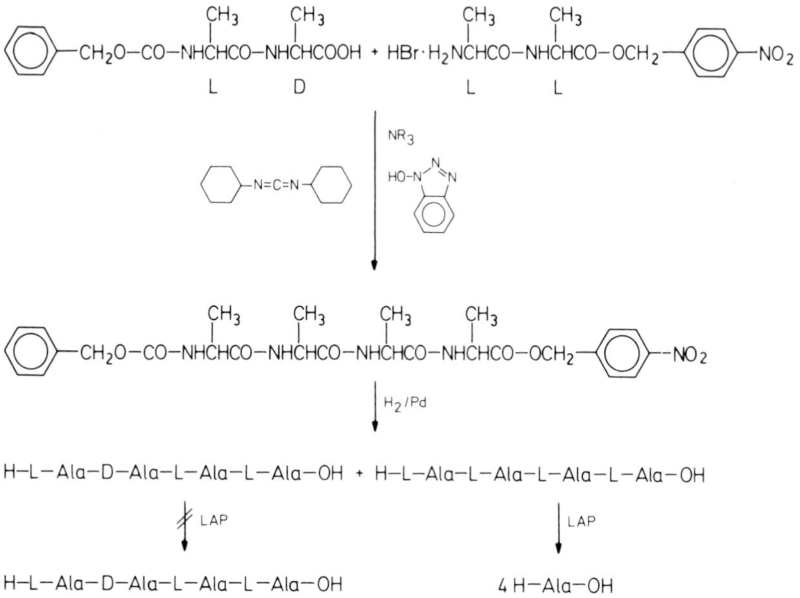

C₆H₅–CH_2O–CO–NHCH(CH_3)CO–NHCH(CH_3)COOH + HBr·H_2NCH(CH_3)CO–NHCH(CH_3)CO–OCH_2–C₆H₄–NO_2

(L, D, L, L)

↓ NR_3, HO–N (1-hydroxybenzotriazole), N=C=N (dicyclohexylcarbodiimide)

C₆H₅–CH_2O–CO–NHCH(CH_3)CO–NHCH(CH_3)CO–NHCH(CH_3)CO–NHCH(CH_3)CO–OCH_2–C₆H₄–NO_2

↓ H_2/Pd

H–L–Ala–D–Ala–L–Ala–L–Ala–OH + H–L–Ala–L–Ala–L–Ala–L–Ala–OH

↓ LAP (no reaction) ↓ LAP

H–L–Ala–D–Ala–L–Ala–L–Ala–OH 4 H–Ala–OH

Benzyloxycarbonyl-L-alanyl-D-alanine [2] (295 mg, 1 mmol) and L-alanyl-L-alanine *p*-nitrobenzyl ester hydrobromide [3] (395 mg, 1.05 mmol) are dissolved in dimethylformamide (6 ml) and 1-hydroxybenzotriazole monohydrate [4] (306 mg, 2 mmol) is added. The mixture is stirred and cooled [5] to $-10\,^{\circ}C$ and a tertiary amine [6] (1 to 2 mmol) is added followed by the addition of dicyclohexylcarbodiimide [7] (206 mg, 1 mmol). The reaction mixture is stirred at $4\,^{\circ}C$ overnight. The precipitated *N*,*N*′-dicyclohexylurea is removed by filtration, washed with dimethylformamide (2 ml) and the filtrate diluted with 0.05 N HCl (160 ml). The solid material is collected on a filter, thoroughly washed with cold water and dried over P_2O_5 in vacuo to constant weight [8].

A sample of the blocked tetrapeptide derivative (114 mg, 200 micromoles) is dissolved in acetic acid (10 ml) and after the addition of a 10% palladium on charcoal catalyst (25 mg) hydrogenated for about four hours. The catalyst is removed by filtration and the solvent evaporated in vacuo. Distilled water (10 ml) is added to the residue and evaporation in vacuo repeated. The residue is suspended in water (0.5 ml) and acetone (10 ml) is added. After storage at 4 °C for about two hours the precipitate is collected, washed on the filter with acetone and dried over P_2O_5 in vacuo [9].

The commercially available suspension of leucine amino peptidase (LAP) [10] is diluted with a 50 mmolar veronal-HCl buffer of pH 8.6 containing $MnSO_4$ (10 micromoles per liter) to a concentration of 50 mg per liter and preincubated at 37 °C one hour prior to its use. A 30 millimolar solution of the substrate (the mixture of two isomeric tetrapeptides, 1.8 mg) in water (0.20 ml) is mixed with the enzyme solution (0.20 ml) in a small test tube and covered with a piece of parafilm. After incubation at 37 ± 1 °C for 3 hours an aliquot (50 microliter) is added to a citrate buffer of pH 2.2 (0.20 ml) and applied to an amino acid analyser [11].

1. Bosshard HR, Schechter I, Berger A (1973) Helv Chim Acta 56: 717
2. Cf. ref. 1. For an alternative method of preparation cf. Erlanger BF, Brand E (1951) J Amer Chem Soc 73: 3508
3. Schechter I, Berger A (1966) Biochemistry 5: 3362
4. The experiment described was used [1] for the determination of the efficiency of 1-hydroxybenzotriazole in the prevention of racemization. In its presence only 0.2% racemization took place while the coupling was accompanied by 80% racemization in its absence. Of course, the influence of other additives can be similarly tested.
5. Racemization is affected by the temperature of the coupling mixture.
6. Triethylamine, *N*-methylmorpholine, diisoprolylethylamine, etc. can be used. The basicity and bulkiness of the tertiary amine has significant effect on the extent of racemization; cf. Bodanszky M, Bodanszky A, Chem Commun 1967: 591. The amount of the base can also be critical.
7. The experiment allows the study of other coupling methods as well.
8. Only if the amount of the crude material is close to the calculated value (0.57 g) can the experiment be considered meaningful. In the case of low recovery a part of the dia-stereoisomer formed in the process might remain in solution and escape detection.
9. The calculated amount for 200 micromole of free peptide is 61 mg. For the weight of the sample to be used in enzymic hydrolysis a correction based on N content (calculated 13.9%) should be applied.
10. In order to ascertain that the enzyme is active and the conditions of the experiment are suitable it is advisable to carry out a parallel hydrolysis of L-Ala-L-Ala-L-Ala-L-Ala as well. For the preparation of Z-L-Ala-L-Ala-OH cf. Stein WH, Moore S, Bergmann M (1944) J Biol Chem 154: 191
11. If complete racemization took place, the sample would contain 1500 nmol L-alanine. Thus 15 nmol alanine found on amino acid analysis correspond to 1% racemization. The sample size should be adjusted to the sensitivity of the amino acid analyzer.

VII Reagents for Peptide Synthesis

1 *tert*-Butyl Azidoformate [1]

tert-Butyl Phenyl Carbonate

$$(CH_3)_3C{-}OH + Cl{-}CO{-}O{-}C_6H_5 \xrightarrow{\text{quinoline}} (CH_3)_3C{-}O{-}CO{-}O{-}C_6H_5 + \text{quinoline} \cdot HCl$$

$C_{11}H_{14}O_3$ (194.2)

A solution of *tert*-butanol (74.1 g = 94 ml, 1 mol) and quinoline (130 g = 119 ml, 1 mol) in dichloromethane (150 ml) is stirred while phenyl chlorocarbonate [2] (157 g = 126 ml, 1 mol) is added dropwise. The rate of addition is regulated to maintain the temperature of the reaction mixture between 38 and 41 °C. After overnight storage at room temperature enough water is added to dissolve the precipitated quinoline hydrochloride. The organic layer is separated and washed twice with water (60 ml each time) and with 5% hydrochloric acid (3 to 4 times, 60 ml each time). The solution is dried over $MgSO_4$, the solvent removed by distillation and the crude *tert*-butyl phenyl carbonate distilled from a Claisen flask. At ca 70 Pa it boils at 74–78 °C [3]. The yield is about 143 g (74%).

tert-Butyl Carbazate

$$(CH_3)_3C{-}O{-}CO{-}O{-}C_6H_5 \xrightarrow{H_2NNH_2} (CH_3)_3C{-}O{-}CO{-}NHNH_2 + HO{-}C_6H_5$$

$C_5H_{12}N_2O_2$ (132.2)

A mixture of the mixed carbonate (97.1 g, 0.50 mol) and hydrazine hydrate (50 g = 48.5 ml, 1.0 mol) is warmed on a hot plate. When the temperature reaches 75 °C spontaneous dissolution occurs with evolution of heat. The temperature will rise to 103 °C. The mixture is allowed to cool to room temperature and stored overnight. Dilution with ether (200 ml) is followed by the addition of a solution of sodium hydroxide (30 g) in water (100 ml) and the two phases are vigorously shaken. The mixture is extracted with ether in a continuous extractor for 24 hours. The ether is removed from the extract by distillation on a steam bath and the residue is distilled from a Claisen flask at 60–61 °C and ca 70 Pa. The purified carbazate weighs about 60 g (90%) [4].

tert-Butyl Azidoformate

$$CH_3-C(CH_3)_2-O-C(=O)-NH-NH_2 + HNO_2 \longrightarrow CH_3-C(CH_3)_2-O-C(=O)-N_3$$

$C_5H_9N_3O_2$ (143.1)

The carbazate (13.2 g, 100 mmol) is dissolved in a mixture of acetic acid (40 ml) and water (80 ml), the solution is stirred and cooled in an ice water bath and treated with $NaNO_2$ (7.6 g, 110 mmol) added in small portions in the course of a few minutes. The reaction mixture is diluted with water (80 ml) and the oily product extracted into ether (twice, 80 ml each time). The extracts are washed with water (80 ml), 1 M $NaHCO_3$ (80 ml) and dried with $MgSO_4$. The ether is removed in the vacuum of a water aspirator at 140–150 mm on a water bath of 40–45 °C. The remaining dark oil, crude *tert*-butyl azidoformate (about 13 g) can be distilled [5] at about 10 kPa and a bath temperature of 90 °C. The fraction which boils at 73–74 °C at 10 kPa weighs 11.5 g (80%).

1. Carpino LA (1957) J Amer Chem Soc 79: 98, 4427
2. Commercially available, usually as phenyl chloroformate. It is toxic, thus the reaction should be carried out under a well ventilated hood.
3. Distillation in the vacuum of a water aspirator leads to decomposition.
4. Extraction of a solution of the product in ether with dilute alkali, redistillation and crystallization from petroleum ether yield a purified material with m.p. of 41–42 °.
5. It would be more advisable, however, to use the crude product without distillation or storage since it can explode. Also, vapors of the azidoformate can cause severe headaches lasting for hours and might produce the symptoms of common cold as well. The operations should be carried out in a well ventilated hood.

2 1-Adamantyl Chlorocarbonate [1]

Adamantyl Chloro-carbonate

1-Adamantyl-OH + $COCl_2$ ⟶ 1-Adamantyl-O–CO–Cl + HCl

$C_{11}H_{15}O_2Cl$ (214.8)

A solution of 1-hydroxyadamantane [2] (15.2 g, 100 mmol) in dry benzene (400 ml) and pyridine (1 ml) is stirred and cooled in an ice-water bath. A solution of phosgene [3] (45 g, ca 450 mmol) in dry benzene (150 ml) is added, dropwise, at about 4 °C for about one hour. When a white solid separates, more benzene is added. After one hour at room temperature the mixture is filtered, the filtrate is poured onto cracked ice, the organic layer separated and dried over anhydrous Na_2SO_4. The solution is concentrated in vacuo to about one-fifth of its original volume and stored in a freezer. For most practical purposes this solution can be used; the yield is assumed to be quantitative. An aliquot of the solution evaporated to dryness in vacuo leaves a residue melting at about 42 °C. Recrystallization from dry petroleum ether (b.p. 30–60 °C) affords the chlorocarbonate in purified form, melting at 46–47 °C.

1. Haas WL, Krumkalns EV, Gerzon K (1966) J Amer Chem Soc 88: 1988
2. Stetter H, Schwarz M, Hirschhorn A (1959) Chem Ber 92: 1629; Commercially available.
3. Preparation of 1-adamantyl chlorocarbonate must be carried out in a well ventilated hood.

3 1-Isobutyloxycarbonyl-2-isobutyloxy-1,2-dihydroquinoline (IIDQ)

IIDQ (coupling reagent)

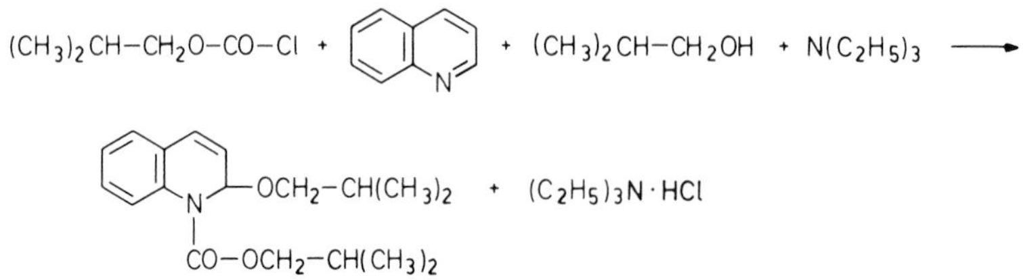

$$(CH_3)_2CH{-}CH_2O{-}CO{-}Cl + \text{quinoline} + (CH_3)_2CH{-}CH_2OH + N(C_2H_5)_3 \longrightarrow$$

$$\text{1-}(CO{-}OCH_2{-}CH(CH_3)_2)\text{-2-}(OCH_2{-}CH(CH_3)_2)\text{-1,2-dihydroquinoline} + (C_2H_5)_3N \cdot HCl$$

$C_{18}H_{25}NO_3$ (303.4)

A mixture of quinoline (12.9 g, 11.8 ml, 100 mmol) and ether (60 ml) is stirred and cooled to $-5\,^\circ C$ and isobutyl chlorocarbonate [2] (15.0 g, 14.3 ml, 110 mmol) is added. Stirring and cooling to $-5\,^\circ C$ are continued while isobutyl alcohol (14.8 g, 18.5 ml, 200 mmol) and triethylamine (11.1 g, 15.5 ml, 110 mmol) are added in small portions. The addition of these two reactants requires about 30 min. The precipitated triethylammonium chloride is removed by filtration and the solvent by evaporation in vacuo. The residue, an oil, is distilled under reduced pressure; the fraction boiling at 144–145 °C at 70 Pa is collected. The product, abbreviated as IIDQ [3], weighs 16.4 g (54%). Its i.r. spectrum shows a characteristic carbonyl band at 1705 cm^{-1}. It gives the expected NMR spectrum in $CDCl_3$ and the calculated values for C, H and N on elemental analysis.

1. Kiso Y, Kai Y, Yajima H (1973) Chem Pharm Bull 21: 2507
2. Commercially available, usually designated as isobutyl chloroformate.
3. This reagent, although an oil, gives somewhat better coupling results than the originally proposed 1-ethyloxycarbonyl-2-ethyloxy-1,2-dihydroquinoline (EEDQ) (Belleau, B, Malek G (1968) J Amer Chem Soc 90: 1651). Both EEDQ and IIDQ can be stored at room temperature for prolonged periods of time.

4 Diphenyl Phosphorazidate (Diphenylphosphoryl Azide, DPPA) [1, 2]

$$(C_6H_5O)_2P(O)Cl + N_3Na \longrightarrow (C_6H_5O)_2P(O)N_3 + NaCl$$

$C_{12}H_{10}N_3O_3P$ (275.2)

Conversion of Carboxylic acids to azides

A mixture of sodium azide (7.8 g, 120 mmol), diphenyl phosphorochloridate [3] (26.9 g = 20.7 ml, 100 mmol) and dry acetone (120 ml) is stirred at room temperature for about a day. The separated sodium chloride is removed by filtration and the solvent by evaporation. The residue is distilled in vacuo. The desired product boils at 152–155 °C and 20 Pa; it weighs 25 g (91%) [4].

Activation of C-terminal Carboxyl group

Coupling reagent

1. Shiori T, Yamada S (1974) Chem Pharm Bull 22: 849; cf. also Scott FL, Riordan R, Morton PO (1962) J Org Chem 27: 4255
2. Diphenyl phosphorazidate converts carboxylic acids into the corresponding acid azides. It can be applied for the activation of the C-terminal carboxyl group in partially protected peptides, and can also be used as coupling reagent if added to a mixture of the carboxyl- and amino-components. Cf. also Yokoyama Y, Shiori T, Yamada S. (1977) Chem Pharm Bull 25: 2423.
3. Commercially available; also as diphenyl chlorophosphate.
4. As an example for the use of diphenylphosphoroazidate the execution of the Young test (cf. p. 183) is described here. A stirred solution of benzoyl-L-leucine (2.35 g, 10 mmol) and glycine ethyl ester hydrochloride (1.53 g, 11 mmol) in dimethylformamide (25 ml) is cooled in an ice-water bath and treated with the reagent (DPPA, 3.03 g = 2.40 ml, 11 mmol). A solution of triethylamine (2.12 g = 2.90 ml, 21 mmol) in dimethylformamide (25 ml) is added dropwise in the course of about 10 minutes. Stirring and cooling with ice-water are continued for about 6 hours. The mixture is diluted with benzene (250 ml) and ethyl acetate (500 ml) and the solution washed with 50 ml portions of N HCl (twice), water, a saturated solution of NaCl in water (twice). The solution is dried over anhydrous Na_2SO_4 and evaporated to dryness in vacuo. The crude product is chromatographed on a column of silica gel (about 500 g). The purified material is eluted with a 10:1 mixture of chloroform and ethyl acetate. Benzoyl-leucyl-glycine ethyl ester thus obtained (2.05 g, 87%) melts at 145–158 °C. The specific rotation of this material, $[\alpha]_D^{20}$ −30.9° (*c* 3, ethanol) indicates that it contains 91% of the L-isomer (excluding the amount present as the racemate). For an isomeric mixture with the same composition Williams and Young (J Chem Soc 1963: 882) report a m.p. of 148–152 °C.

5 3-Hydroxy-3,4-dihydroquinazoline-4-one (4-Hydroxyquinazoline 3-oxide) [1, 2]

$C_8H_6N_2O_2$ (162.1)

2-Aminobenzhydroxamic Acid (precursor)

To a solution of sodium hydroxide (16 g, 400 mmol) in water (100 ml) hydroxylamine hydrochloride (13.9 g, 200 mmol) is added in small portions with stirring. Methyl anthranilate (15.1 g = 13.0 ml, 100 mmol) is added followed by enough methanol to bring it into solution. After three days at room temperature the solution is concentrated in vacuo to about 30 ml, the material which separates, the sodium salt of 2-aminobenzhydroxamic acid, is collected on a filter and washed with ether. The aqueous filtrate is acidified to Congo with 6 N hydrochloric acid, the free hydroxamic acid which precipitates is collected, washed with water and dried in air. Extraction of the crude material with warm ether in a Soxhlet extractor yields the purified acid. It is light brown, and melts at 149 °C. The combined yield of the sodium salt and the free acid is about 70%.

3-Hydroxy-3,4-dihydroquinazoline-4-one

o-Aminobenzhydroxamic acid (1.52 g, 10 mmol) and 98% formic acid (3 ml) are heated to boiling under reflux for 15 min. Water (10 ml) is added and boiling is continued for an additional 10 min. On cooling the product separates. It is collected on a filter, washed with water and dried in air; it weighs about 1.5 g (92%) and melts at 242–244 °C.

1. Harrison D, Smith ACB, J Chem Soc. 1960: 2157.
2. This compound is an efficient catalyst in acylation with active esters; cf. König W, Geiger R (1973) Chem Ber 106: 3626
3. Scott AW, Wood BL, Jr (1942) J Org Chem 7: 508

6 1-Hydroxybenzotriazole [1, 2]

Hydroxybenzotriazole

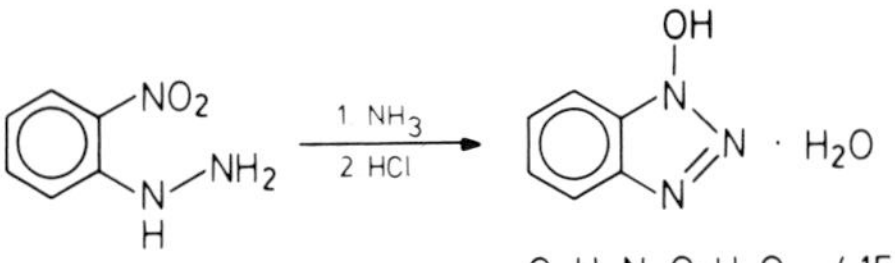

$C_6H_5N_3O \cdot H_2O$ (153.1)

To a suspension of *o*-nitrophenylhydrazine (15.3 g, 100 mmol) in water (100 ml) concentrated ammonium hydroxide (100 ml) is added and the mixture is heated on a steam bath with occasional swirling. The starting material dissolves and the color of the solution changes from dark red to light brown. Heating to about 70 °C is continued for 15 min and the solution is concentrated on the steam bath with the help of an air stream to about half of its original volume. After filtration from a small amount of insoluble material the solution is acidified to Congo with 6 N HCl. The mixture is cooled in an ice-water bath for about 30 min, the crystals collected on a filter, washed with cold water (30 ml) and dried in air. The product (11 g, 72%) melts at 154–156 °C [3]. It is sufficiently pure for most practical purposes, but the m.p. can be raised to 157 °C by recrystallization from dilute ethanol.

1. Nietzki R, Braunschweig E (1894) Ber dtsch Chem Ges 27: 3381

2. 1-Hydroxybenzotriazole is commercially available as the monohydrate.
3. At about 100 °C there is a change in the appearance of the solid.

7 Ethyl 2-hydroximino-2-cyanoacetate [1, 2]

Ethyl hydroximino cyanoacetate

$$NC{-}CH_2{-}CO{-}OC_2H_5 + NaNO_2 + CH_3COOH \longrightarrow NC{-}\underset{\displaystyle NONa}{\underset{\|}{C}}{-}CO{-}OC_2H_5 \xrightarrow{HCl} NC{-}\underset{\displaystyle NOH}{\underset{\|}{C}}{-}CO{-}OC_2H_5$$

$C_5H_6N_2O_3$
(142.1)

Ethyl cyanoacetate (11.3 g, 100 mmol) is added to a solution of sodium nitrite (8.3 g, 120 mmol) in distilled water (50 ml) and acetic acid (8.0 ml = 8.4 g, 140 mmol) is added to the stirred mixture. The ester disappears and soon yellow crystals of the sodium derivative start to separate. Next day the crystals are collected and then dissolved in 2 N HCl (50 ml). The product is extracted with ether (four times, 50 ml each time) and the extracts are dried over anhydrous Na_2SO_4. Removal of the solvent by evaporation in vacuo leaves a crystalline residue melting at 133 °C [3]. (12.4 g, 87%).

1. Conrad M, Schulze A (1909) Ber dtsch Chem Ges 42: 735
2. For the prevention of racemization with this reagent (and also with the 2-hydroximino-2-cyanoacetamide) cf. Itoh M (1943) Bull Chem Soc Jpn 46: 2219
3. Ethyl 2-hydroximino-2-cyano acetate can be purified by recrystallization from boiling water.

8 1-Guanyl-3,5-dimethyl-pyrazole Nitrate [1, 2]

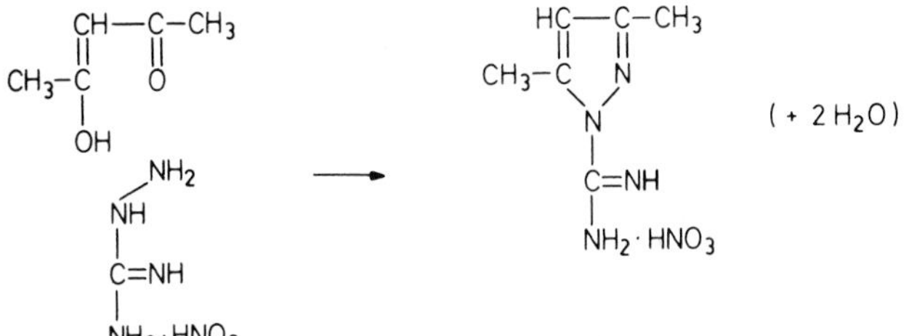

Guanylation

$$CH_3-C(OH)=CH-C(=O)-CH_3 + NH_2-NH-C(=NH)-NH_2 \cdot HNO_3 \longrightarrow \text{1-guanyl-3,5-dimethyl-pyrazole} \cdot HNO_3 \; (+ 2\,H_2O)$$

$C_6H_{11}N_5O_3$ (201.2)

Reagent for converting ornithine residues to arginine residues

A mixture of 2,4-pentanedione (10 g = 10.3 ml, 100 mmol), ethanol (15 ml) and water (15 ml) is heated under reflux and aminoguanidine nitrate (13.7 g, 100 mmol) is added in small portions. The solution is heated to reflux for two more hours, then allowed to cool to room temperature. Next day the crystals which separated are collected on a filter, washed with ether and dried in air. Addition of ether to the filtrate yields additional crops. The crude material (about 16 g) is recrystallized from hot ethanol; 1-guanyl-3,5-dimethyl-pyrazol nitrate (12.9 g, 64%) melts at 166–168 °C.

1. Bannard RAB, Casselman AA, Cockburn WF, Brown GM (1958) Canad J Chem 3, 76: 1541; cf. also Thiele J, Dralle E (1898) Liebigs Ann Chem 302: 275
2. 1-Guanyl-3,5-dimethyl-pyrazole reacts with primary amines to form guanidine derivatives (and 3,5-dimethyl-pyrazole). It has been recommended for the guanylation of the side chain amino groups of lysine residues in proteins (Habeeb AFSA (1960) Canad J Biochem Physiol 38: 493) and is also suitable for the conversion of ornithine residues to arginine residues in synthetic peptides (Bodanszky M, Ondetti MA, Birkhimer CA, Thomas PL (1964) J Amer Chem Soc 86: 4452).

VIII Appendix

1 Conversion of Dicyclohexylammonium Salts of Protected Amino Acids to the Free Acids [1]

N-o-Nitrobenzensulfonyl-O-tert-butyl-L-threonine-N-Hydroxysuccinimide Ester

$C_{18}H_{23}N_3O_7S$ (425.5)

The finely powdered dicyclohexylammonium salt of the protected amino acid, (5.1 g, 10 mmol) is added to a two phase system of ethyl acetate (40 ml) and a solution of $KHSO_4$ (2 g, 15 mmol) in water (40 ml) [2]. The mixture is shaken by hand until the dicyclohexylammonium salt is completely dissolved. The aqueous phase is extracted with ethyl acetate (twice, with 10 ml each time). The combined organic layers are washed with water until they are free of sulfate ions, then dried over anhydrous sodium sulfate and evaporated to dryness in vacuo [3].

The residue is dried in vacuo, over P_2O_5 for several hours; the protected amino acid is dissolved in tetrahydrofurane (60 ml) and *N*-hydroxysuccinimide [4] (1.16 g, 10 mmol) is added. The solution is then cooled to about 0 °C, and dicyclohexylcarbodiimide (2.06 g, 10 mmol) is added with stirring. After two days at this temperature the *N,N'*-dicyclohexylurea is removed by filtration and washed with tetrahydrofurane. The filtrate is evaporated in vacuo, the residue triturated with ethanol (60 ml) to induce crystallization. The crystals are collected, dried over P_2O_5 in vacuo (1.3 Pa) at 30 °C. The active ester, (3.56 g, 83%), melts at 138–140 °C, $[\alpha]_{546}^{20}$: −56° (*c* 1, dimethylformamide).

1. Spannenberg R, Thamm P and Wünsch E (1971) Hoppe-Seyler's Z Physiol Chem 352: 655
2. It seems to be advantageous to replace the conventionally used aqueous citric acid solutions by a 5% solution of potassium hydrogen sulfate ($KHSO_4$) in water. This circumvents the risk of dissolving some citric acid in the reaction mixture.
3. If the active ester can be prepared in EtOAc, this dried solution can be used in the esterification step.
4. See footnote 3, on p. 105

2 Preparation of Analytical Samples

2.1 Elemental Analysis

Over and above the usual requirements such as the elimination of chemical and mechanical impurities by recrystallization or reprecipitation [1] peptides have to be dried with exceptional care. Free peptides, often lyophilized materials, retain moisture tenaciously. In addition to water they might contain acetic acid as well. In order to obtain reproducible values the samples have to be dried in vacuo, at elevated temperature, for prolonged periods of time. The temperature of drying should be based on the melting point or decomposition point of the material. While it is desirable to choose a temperature as high as 110 °C or even 130 °C, drying should not be accompanied by melting or decomposition.

Protected peptides are usually insoluble in water. Nevertheless, they can be hygroscopic in a sense: they might absorb water from moist air. Hence, for the sake of valid analyses such samples must be dried at 110 to 130 °C and allowed to cool in a desiccator in the presence of phosphorus pentoxide. In particularly difficult cases the sample should be weighed and dried, by the microanalyst, in a "piggy", which is closed while still hot and reweighed after cooling.

Probably because of the presence of weak basic centers, such as nitroguanidines, the imidazole of histidine or the amide groups themselves, protected peptides can contain acetic acid or trifluoroacetic acid, even after drying. This possibility should be considered at the evaluation of the value of elemental analysis. For best information all the elements present in the compound should be determined and the sum should amount to 100%.

Lengthy considerations and experience taught the authors to value the results of combustion analysis. Spectral methods should complement but not replace the determination of elemental composition.

2.2 Amino Acid Analysis

If the synthetic material is available in sufficient amounts, a 3–5 mg sample should be used and weighed with an accuracy of 0.1 mg. This allows the calculation of "recovery" from the values of amino acid analysis and thus the determination of the peptide content of the analyzed substance. The peptide content of a sample can be far below 100% since peptides, both free and protected, can retain significant amounts of moisture and/or non-volatile inorganic impurities.

Most commonly, constant boiling hydrochloric acid is used for hydrolysis. This is prepared by diluting concentrated hydrochloric acid with an equal volume of water and distilling the mixture from an all-glass apparatus. The first part of the distillate about 10% is discarded and about 10% of the acid is left undistilled. The distillate, about 5.7 N HCl, is stored in several small glass-stoppered bottles or in sealed ampoules to prevent the absorption of ammonia from the air during frequent opening.

The sample is weighed into a soft glass ampoule with a narrow neck and dissolved in a small volume (e.g. 0.5 ml) of constantly boiling hydrochloric acid. The solution is cooled in an ice-water bath, evacuated with the help of a water aspirator and sealed under vacuum by melting the narrowed part of the tube with a small Bunsen burner. The sealed ampoule is placed into a cavity of an electrically heated metal block and kept at 110 °C for 16 hours.

Peptides containing valine and/or isoleucine residues [2] as next neighbors in their sequence require longer periods for complete hydrolysis, sometimes as long as 3–4 days. On long hydrolysis, however, significant decomposition of serine and threonine takes place. Therefore, for a precise determination of the amino acid ratios both short and long hydrolyses are necessary. From several analyses, e.g. after hydrolysis for 16, 32 and 64 hours, the true values for valine and isoleucine can be found by graphical extrapolation to infinite time while extrapolation to zero hour gives the hypothetical amount of serine and threonine in the hydrolysate in the absence of decomposition [3].

Tryptophan is destroyed in hot hydrochloric acid, particularly in the presence of air and heavy metal impurities. With highly purified hydrochloric acid, tryptophan-containing peptides can be hydrolyzed in vacuo and the tryptophan content determined [4]. Hydrolysis with 3 molar β-mercapto-ethanesulfonic acid or 4 molar methanesulfonic acid causes no significant decomposition of the indole. Methionine sulfoxide is converted in part to methionine during hydrolysis with hydrochloric acid. The conversion is complete if β-mercaptoethanol (1 mg per ml) is added. The sulfoxide remains intact during hydrolysis with 3 N *p*-toluenesulfonic acid.

On completion of the hydrolysis the sealed ampoules are cooled to room temperature and carefully [5] opened. The hydrolysate is transferred with a pipet into a small beaker, the ampoule rinsed with distilled water and the acid is removed on a steam bath with the help of a stream of nitrogen. A small volume of distilled water is added and similarly evaporated. The residue is dissolved in a buffer of pH 2.2 and, after appropriate dilution, applied to the column of the amino acid analyzer. The applied volume depends upon the kind of instrument used. The buffer, glass vessels, pipet, etc. must be clean, especially if a sensitive method of analysis involving samples of 10 nanomoles or less is applied. Several amino acids, aspartic acid, serine and glycine particularly, are present in fingerprints in amounts which can distort the results of amino acid analyses.

2.3 NMR Spectra

A tendency for solvent retention often complicates the recording of NMR spectra. It is advisable, therefore, to dissolve the (dried) sample in the deuterated solvent, to evaporate the solvent in vacuo and to redissolve the residue.

The most commonly used solvent in NMR spectroscopy, $CDCl_3$, has limited applicability in peptide chemistry. Only relatively small molecules such as protected and activated amino acids and small peptides are soluble in chloroform. Deuterated dimethylformamide, $(CD_3)_2NCDO$, although expensive, is more generally useful, as is deuterio-dimethylsulfoxide. Very good results were obtained in the authors' laboratory with fully deuterated acetic acid, CD_3COOD. The latter, unlike trifluoroacetic acid, which is also used in NMR spectroscopy of peptides, leaves most protecting groups intact. Also, the chemical shifts in CD_3COOD are quite close to those recorded in $CDCl_3$ and thus allow a better comparison with chemical shifts reported in the literature. Trifluoroacetic acid has a major effect on the chemical shifts and renders the use of published values rather difficult. Last, but not least, in CD_3COOD exchangable protons on oxygen and nitrogen atoms are displaced by deuterium. This results in a welcome simplification of the spectra. To bring the exchange to completion the samples are dissolved in CD_3COOD, the solvent removed by evaporation [6], the residue redissolved, and the evaporation repeated. Finally, the solution of the residue in CD_3COOD is applied for the recording of the spectra. The small amount of CHD_2COOD present in the solvent can serve as internal reference.

1. Amino acids and their derivatives can generally be sublimed in vacuo. In the authors' laboratory a cyclo-pentapeptide was sublimed without any decomposition at about 270 °C and 4 Pa.
2. Some residues with blocked functional side chains, e.g. *S*-benzyl-cysteine, have similar influence on the rate of the hydrolysis of the peptide bonds surrounding them.
3. Some protected peptides remain insoluble in hot constant boiling hydrochloric acid. A mixture of equal volumes of concentrated hydrochloric and acetic acid can be helpful in such instances. It should be noted, however, that hydrolysis with this mixture is conducive to racemization.
4. The most practical method for the determination of the tryptophan content of a peptide is the recording of the U.V. spectrum. Tryptophan has a molar absorption coefficient of 5500 at 280 nm. Only tyrosine absorbes in the same order of magnitude (1370 at 278 nm). Phenylalanine (about 200 at 260 nm) interferes less with this determination, but nitroarginine has too high molar absorption to allow the exact determination of tryptophan content.
5. The eyes must be protected during breaking of the sealed part of the ampoule.
6. The solvent can be removed by lyophilization (sublimation from the frozen solution) as well.

Subject Index